X-ray Astronomy

X-ray Astronomy

J. LEONARD CULHANE
and PETER W. SANFORD

FABER AND FABER
London · Boston

First published in 1981
by Faber and Faber Limited
3 Queen Square London, WC1N 3AU
Printed in the United States of America by Arrangement
with Charles Scribner's Sons All rights reserved

British Library Cataloguing in Publication Data

Culhane, J Leonard
X-ray astronomy.
1. X-ray astronomy
I. Title   II. Sanford, Peter W
522'.686   QB472

ISBN 0–571–11550–0

Contents

Acknowledgements

We are grateful to the following workers and their publishers who have provided material for the illustrations. Plate 1: Robert McQueen, High Altitude Observatory, Colorado; plate 2: Fred Hawkins, OAO Corporation MD USA; plate 3: Frank Malina; plate 4: British Aerospace Dynamics Group; plate 5: Appleton Laboratory of the UK Science Research Council; plate 7: Al Krieger, American Science and Engineering Inc.; plate 8: Alan Title, Lockheed Palo Alto Laboratories; plate 10: adapted from a composite by T. A. Mathews; plates 12 and 18: Royal Astronomical Society; plate 15: Martin Ryle, Cambridge, Roger Lynds, Kitt Peak and Paul Gorenstein, Harvard; plates 16 and 17: Royal Observatory, Edinburgh.

For the line figures: fig 1.1: calculations by Giovanni Bibbo (Space Science Department ESA); fig 1.2: from an article by Leo Goldberg 'Ultraviolet Astronomy' *Scientific American* June 1969; fig 2.1: Lucio Chiapetti and Derek Hoyle, Mullard Space Science Laboratory, University College, London; fig 3.2a: from the work of Loren Acton, Jake Wolfson (Lockheed Palo Alto Laboratories); fig 3.2b: from the work of John Parkinson, (Mullard Space Science Laboratory); fig 3.2c: from the work of George Doschek (US Naval Research Laboratory); fig 4.3: Simon Mitton, from *Exploring the Galaxies*, Faber & Faber; fig 4.4: Peter Sturrock (Stamford University); fig 5.1: R. Giacconi and colleagues, *Astrophysical Journal* letters 168 L27, University of Chicago Press; figs 7.2, 9.1: from *Neutron Stars, Black Holes and Binary X-ray Sources*, H. Gursky and R. Ruffini, D. Reidel, Dordrecht; fig 7.3: from Ed Van den Heuvel's paper in the Eighth Texas Symposium in Relativistic Astrophysics, New York Academy of Sciences (1977); fig 8.2: from the work of Saul Rappaport; fig 10.1: from the work of Martin Rees, Joseph Silk and Michael Rowan Robinson; fig 12.2: from G. C. Pooley and Martin Ryle, Monthly Notices of the Royal Astronomical Society *139* 515 (1968); fig 12.5: from a drawing by Kim Ward, Appleton Laboratory of the UK Science Research Council.

Illustrations

Figures

Authors' Preface

X-ray astronomy has been established as a powerful means of observing matter in its most extreme form. The energy liberated by sources discovered in our Galaxy has confirmed that collapsed stars of great density, and with intense gravitational fields, can be studied by making observations in the X-ray part of the electromagnetic spectrum. The astronomical objects which emit detectable X-rays include our own Sun and extend to quasars at the edge of the Universe.

This book describes the history, techniques and results obtained in the first twenty-five years of exploration. Space rockets and satellites are essential for carrying the instruments above the Earth's atmosphere where it becomes possible to view the X-rays from stars and nebulae. However, the foresight of scientists such as Bruno Rossi, Riccardo Giacconi and Herbert Friedman in the United States, and Harrie Massey and Robert Boyd in the United Kingdom was crucial in promoting the rapid growth of this new astronomy. Without their inspiration and the encouragement of many administrators in the American National Aeronautics and Space Administration (NASA) and the British Science Research Council (SRC) and in the space agencies of other governments the subject could not have achieved the maturity that has allowed us to record the remarkable discoveries about objects which range from black holes to the largest systems of the Universe—the clusters of galaxies.

Many colleagues assisted us whilst writing this book. We thank, in particular, Professor Robert Boyd and Simon Mitton for their constructive comments. The expertise of Jane Salton, Libby Daghorn and Derek Hoyle in preparing the manuscript is acknowledged with gratitude.

September 1980   J. LEONARD CULHANE
                                    PETER W. SANFORD

1 · The birth of X-ray astronomy

1.1 The new astronomy

Astronomy, the ancient science available to all who think deeply and are able to look to the heavens, has been transformed in the twenty-five years since the early 1950s. Up to that time, indeed during the 350 years that elapsed after the telescope was first used by Galileo, progress depended mainly on the construction of larger telescopes in more favourable locations—usually on the tops of mountains. These improvements provided greater collecting power and better seeing conditions, which enabled the study of fainter and more distant objects in the Universe.

The revelations of the last twenty-five years came when radiation from the stars invisible to the eye was detected and measured. The new methods of observation, using radio waves, infrared, ultra-violet, X-radiation and gamma rays, have disclosed a new vision of the Universe.

The best known and most important example of the contributions made with these new methods was the discovery, by Penzias and Wilson in 1965, of a background of radio waves in the microwave region of the spectrum. This radiation had been predicted in George Gamow's theory that the Universe began with a mighty explosion known as the 'Big Bang'. A faint background of microwave radiation was the predicted remnant of the initial fireball.

Measurement of a significant portion of the radio spectrum can be made with arrays of aerials even at sea level: the famous array of telescopes of the Cavendish Laboratory at Cambridge provides an excellent example. However, optical, and most of the other techniques of astronomy, are best conducted—sometimes must be conducted—above the effects of the Earth's atmosphere. The recent history of astronomy is thus intimately connected with man's achievements in vertical ascent.

In 1910, the Eiffel Tower, at that time the highest man-made structure, was used to test the effects of elevation on the measurement of ionizing radiation, which had recently been discovered. An electroscope was carried to the top of the tower by Wulf, who found that its rate of discharge increased with altitude. The discharge of an electroscope was known to be more rapid in the presence of ionizing radiation from radioactive substances. The slow discharge that always occurred, even in the absence of known radioactivity, indicated that a low level of radiation was present at the Earth's surface. When the experiment was repeated, in balloon flights of up to six miles in altitude, it was established conclusively that the ir.creased radiation was of cosmic origin and not entirely due to the natural radioactivity of the Earth's crust.

Ionizing radiation, which we now know is caused by very energetic particles called cosmic rays incident on the atmosphere, is best studied with balloons, rockets and artificial satellites. The characteristics of the primary cosmic rays can be determined with precision only before they start to interact with the atmosphere, where they produce a shower of secondary particles.

X-radiation—in common with cosmic rays—also produces ionization, which can discharge an electroscope. For this reason there was considerable uncertainty about the true nature of X-rays when they were discovered by Röntgen in 1895. He called them X-rays to indicate their unknown nature. As we shall see in Chapter 2, the ability to produce ionization is one of the few properties they have in common with cosmic rays.

We are used to thinking of X-rays as being more penetrating than visible light. They are, for example, used extensively in the investigation of the internal structure of materials, and their use in medicine for the investigation of bone fractures and internal organs began shortly after the radiation was discovered. We might, therefore, be surprised how rapidly the atmosphere absorbs X-rays—a few millimetres of air is sufficient to absorb low-energy (1 keV) X-rays for example. However, if we remember that the mass of air above the surface of the Earth is sufficient to support a column of mercury in a barometer to a height of 0.76 metres, then we can begin to understand the reason. A few centimetres of mercury will absorb X-rays with energy as high as 100 keV. The penetration of X-rays incident on the Earth's surface is illustrated in Fig. 1.1, which shows the height at which the incident radiation is reduced to 50 per cent by absorption. The depth of penetration depends strongly on the energy of the X-rays, which is measured in keV (thousands of electron volts). This unit of measurement is generally used in the laboratory where X-rays are produced by electrons made to accelerate with a high voltage before they strike a metal anode. The maximum energy of the radiation produced is given by the applied voltage.

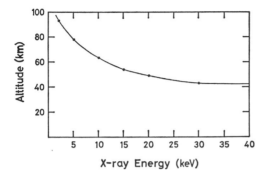

Fig. 1.1 The penetration of X-radiation into the Earth's atmosphere. The curve illustrates the altitude at which 50 per cent of the X-rays incident vertically on the Earth's atmosphere is absorbed. The lower energy X-rays are absorbed at higher altitudes

We can immediately see from Fig. 1.1 that it is impossible to detect X-rays from the Sun or the stars unless the instrument can be taken to altitudes of 100 km for the low energies and more than 30 km for the high energies. The liquid-fuelled rocket developed during the 1940s provided the means of discovering X-radiation from space.

1.2 The high-altitude rocket

As a military weapon, the rocket has a history dating back to AD 1200; however, its use as a research vehicle is quite recent and started with the development of the liquid-propellant rocket. The theory behind the operation of the rocket was worked out in detail by an American, Robert Goddard, and published in 1919 at the time when he was a young Professor of Physics at Clark University. Goddard saw the potential of the rocket as a research vehicle capable of achieving higher altitudes than balloons. His contribution to the origins of space science are now recognized in a research centre of the National Aeronautics and Space Administration (NASA) which is located near Washington, DC, and has been named the Goddard Space Flight Center.

In Germany, Hermann Oberth also developed the theory of rocket propulsion and by 1933 a number of small liquid-propellant rockets had been fired by the Verein für Raumschiffahrt (Society for Space Flight). Some of the rockets went as high as one mile. A detailed analysis of the flight performance for a rocket powered by successive impulses was published in 1938 by H. S. Tsien and Frank Malina working at the Californian Institute of Technology (CALTEC) in a group led by Theodore Von Kármán. The now famous Jet Propulsion Laboratory (JPL)* became a centre of research on long-range missiles and space exploration with Frank Malina as its first director.

At the end of the Second World War, scientists at JPL developed the first high-altitude rocket primarily for meteorological research. The WAC Corporal (see Plate 3) became, in the autumn of 1945, the first high-altitude rocket to be used for space research. In its first flight, on 11 October 1945, it reached an altitude of over forty miles. This early success eventually gave birth to other research rockets— notably the Aerobee which continued until at least 1980, providing a vehicle with which scientists could study the physics of the upper atmosphere and make astronomical observations from above its obscuring influence.

A high-altitude rocket was also developed in Germany during the Second World War. It was a 14-ton guided missile called V-2 and was first used in the spring of 1942. However, it was built for range, not altitude; to achieve a range of 190 miles it had to reach an altitude of sixty miles. By the end of the war some 3,000 of these missiles had been produced and over 1,000 had reached English soil.

Early in 1945, the Allies captured a large number of the V-2 rockets, and in January 1946 the United States Ordnance Department invited scientific groups to use the missiles to conduct experiments at high altitudes. By October 1951 sixty-six V-2s had been launched from the White Sands Proving Ground in New Mexico. More than 20 tons of scientific equipment had been carried to altitudes of up to 250 miles.

Many organizations played a part in the exploration of the upper atmosphere made possible with the V-2 rockets. The US Naval Research Laboratory (NRL) supplied instrumentation to replace the warhead, both for other agencies and for its own research pro-gramme. By 1948, NRL scientists had made the first detection of X-rays from the Sun with photographic plates covered by thin

* An account of the Laboratory's activities from 1943 to 1946 can be obtained in a NASA report entitled *Essays on the History of Rocketry and Astronautics*, 1977.

beryllium filters. This experiment, by Burnight, marked the beginning of X-ray astronomy, but the effects of the Sun's radiation on the upper atmosphere were of more concern at that time than the solar conditions which could produce such energetic radiation.

The electrified layers of the outer atmosphere of the Earth were a particularly high priority for study, for it was known that they provided the means for long-distance radio communication. Heaviside and Kennelly had independently suggested in 1902 that the upper atmosphere contained a conducting layer which could, by multiple reflections, explain the bending of radio waves around the curved surface of the Earth. The existence of these layers, now known as the ionosphere, was demonstrated in 1925 by Appleton and Barnett in England and Breit and Tuve in the United States.

High-altitude rockets made it possible both to measure the conditions in the ionosphere and also to measure the solar radiation that produces the ionization. A significant technical advance was made by Friedman and his colleagues, who introduced the photon counter (see Chapter 2) to measure the energetic solar radiation. This device, a Geiger counter which responds to ultraviolet and X-radiation, allowed continuous measurements throughout the flight of the rocket. By 1949 Friedman had demonstrated that X-rays from the Sun were at least as energetic as 2 keV and that they could therefore produce ionization in the lower part of the ionosphere.

Hulbert and Vegard (see p. 46) had predicted that the Sun should emit X-radiation. When eclipsed by the Moon, the diffuse, less luminous outer region, known as the corona, becomes visible. This region can be seen in the beautiful photograph (Plate 1) obtained from an eclipse expedition by the University of Colorado on 12 November 1966. The Moon provides an opaque disc, which completely obscures the bright light of the photosphere. The corona emits a continuous spectrum and a number of discrete lines in the visible region. The identification of these emission lines remained a puzzle for many years until in 1942 Grotrian and Edlén demonstrated that they arose from atoms that had been stripped of a large number of their electrons. The temperature of the corona, implied by the existence of these highly ionized atoms, had to be at least 1,000,000 K (degrees Kelvin). Such a temperature produces radiation in the X-ray region of the spectrum. As we shall see in Chapters 3 and 4, the studies of the Sun's X-ray emission are now very detailed and this region of the Sun has provided a natural laboratory where the properties of gases at very high temperatures can be explored.

The early success of the V-2 research programme led to the development of rockets more suitable for continuing the studies of the upper atmosphere and eventually for making detailed studies of the Sun and the stars. In the United States, the Aerobee rocket provided the means for continuing the research after the supply of V-2 rockets came to an end in 1952. The Veronique rocket became available in the French programme of upper atmospheric research, and in the United Kingdom, a solid-fuelled rocket called Skylark, conceived in 1953 after a conference held in Oxford, revealed the richness of the phenomena that could be observed with high-altitude rockets.

By 1957, programmes of research using sounding rockets had been established in several countries. However in October of that year the first artificial satellite was launched from the USSR, an event which shook the world but allowed astronomers to realize their dreams of telescopes in orbit. On 1 October 1958, the United States Congress established NASA with the declared policy that activities in space should be devoted to peaceful purposes for the benefit of all mankind. The achievements of both NASA and the USSR in space exploration during the decade of the 1960s were certainly dramatic, and although there was more emphasis on the provision of space spectacles than space telescopes, space astronomy also flourished through a healthy programme of research satellites.

1.3 Orbiting observatories

Soon after the formation of NASA, three programmes of space exploration were started, each using a specially designed satellite. The orbiting geophysical observatory (OGO) provided a means for studying the environment of the Earth, the complexities of the interaction of the particles flowing from the Sun with the Earth's magnetic field. The orbiting solar observatory (OSO) provided a continuous watch of the activity on the Sun by making detailed measurements of radiation emitted from the turbulent outer atmosphere of the Sun. The third programme consisted of using the orbiting astronomical observatory (OAO), which was designed to provide a stable platform from which observations of the stars could be made.

The provision of these three satellites was an imaginative concept and, in the case of the OSO, particularly cost effective since the satellite was a standardized, and at the same time a very elegant, solar observatory. Fig. 1.2 shows how the satellite was constructed. The upper part of the satellite, which includes the solar array and usually two instrument boxes, was continuously pointed at the Sun. The lower part, a structure like a spinning wheel, contained a number of similar compartments which could also accommodate instruments and these could view the Sun as it passed with each revolution of the wheel. In keeping with the objectives of NASA, many instruments from other countries were given flight opportunities on these OSO spacecraft.

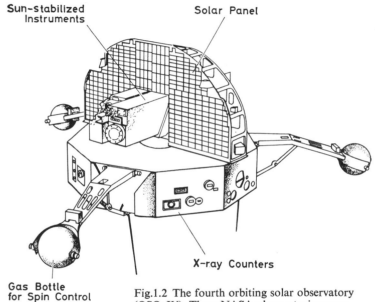

Sun-stabilized Instruments

Solar Panel

X-ray Counters

Gas Bottle for Spin Control

Fig.1.2 The fourth orbiting solar observatory (OSO–IV). These NASA observatories were used in the 1960s and 1970s to study both the Sun and the stars. Instruments in the rotating wheel section could scan the sky in the search for radiation for stars

The exploratory studies of the Sun with high-altitude rockets had already shown the importance of making measurements in the ultraviolet and X-ray regions of the spectrum. Instruments covering these regions were therefore selected, with priority, for flight at the beginning of the OSO programme. At first, the X-ray instruments were similar to those which had been used in the rocket programme. Later on, the simple photon counters of the early satellites were replaced with instruments that could make detailed measurements of the spectrum or provide the locations and structures of the X-ray-emitting regions on the Sun's disc. The evolution of instrumentation in this programme was to provide a sound technology for the more exacting requirements which were to come with the discovery that X-rays could be detected from astronomical objects other than the Sun.

The Sun is an intense source of X-radiation outside the Earth's atmosphere. However, if it were to be placed at the distance of the nearest star (Proxima Centauris), the flux of radiation would be minute. The reason for this is that the radiation spreads out over a sphere so that the intensity decreases rapidly with distance according to the inverse-square law. Proxima Centauris is 150,000 times more distant from the Earth than the Sun so that any signal originating from that much further away would be diminished by the staggering factor of 0.000 000 000 04, or 4×10^{-11} for brevity.* The normal X-ray flux from the Sun, at the distance of the Earth, is about 1,000,000 photons per second falling on each square centimetre. We might therefore expect about three photons every day per square centimetre from Proxima Centauris if it has an X-ray-emitting corona similar to that of the Sun. This flux value would normally be written in the form 4.0×10^{-5} photons/sec/cm^2. The detection of such a small signal presents considerable difficulties and for this reason there could be no certainty that X-rays would ever be detected from stars other than the Sun.

These pessimistic estimates did not deter the pioneers of X-ray astronomy; two considerations made the search for other X-ray-emitting stars worthwhile. First the Sun sometimes flares to increase its X-ray emission by more than one thousandfold, and second the Sun is a relatively inactive star. There are many stars that are known to have more violent activity in their atmospheres, and there existed the possibility of detecting more intense coronal activity in such a nearby star. With these considerations in mind, scientists in the United Kingdom, led by Robert Boyd of University College, London, proposed X-ray instruments for flight on NASA's orbiting astronomical observatory.

The OAOs were to be used primarily for the study of the ultraviolet emission of stars. Unlike X-ray astronomy the scientific value of such observations was not in question. All wavelengths shorter than about 300 nanometers (nm) are blocked out by the atmosphere,

---

* Astronomers abbreviate large and small numbers either by use of logarithms or writing to the powers of 10 as in this example.

but much valuable information can be given by the spectra of stars between this atmospheric limit and the cut-off at 90 nm, which is caused by hydrogen in the space between the stars. We have already mentioned that the Sun has a diffuse outer region called the corona, which radiates X-rays and ultraviolet, but most of the energy from the Sun comes from the region which we see as a disc and which is known as the photosphere. The effective temperature of the Sun is only 6,000 K because the much hotter corona contributes about one millionth (10^{-6}) of the total energy output. Most of the radiation from the Sun falls in the visible range of the spectrum or at longer wavelengths such as the infrared and radio regions. When the temperature of a star is as high as 25,000 K the peak of the emission spectrum is at shorter wavelengths and 80 per cent of this ultraviolet radiation is absorbed by the Earth's atmosphere. The OAO satellites made possible detailed measurement of the radiation from such stars. In addition to providing the characteristics of the atmospheres of these hot stars, the ultraviolet measurements also provide valuable data on the diffuse material between the stars. It is in this region of the spectrum that the effects of this material can be observed most easily in the form of very narrow absorption lines.

The University College instruments designed to study X-rays from stars were flown on the last OAO, which was named Copernicus in honour of the Polish astronomer, born in 1473, who had caused the revolution in man's view of his position in the Universe by pointing out that the Earth was not at the centre but merely in orbit about the Sun. The satellite carries an ultraviolet spectrometer of great precision from Princeton University Observatory, and this has made an extensive study of the interstellar medium—the diffuse material between the stars which gives narrow absorption lines in the spectra of the hot stars mentioned above. The X-ray telescopes are pointed in the same direction as the ultraviolet telescope so that simultaneous measurements can be made of any X-rays also coming from the stars used in the Princeton investigations. A photograph of the satellite can be seen in Plate 2. The satellite was launched in August 1972 and, as we shall see, the initial objectives of the X-ray instruments were completely redirected by the discoveries that had been made after the proposal was first put.

In spite of the dramatic events of the intervening ten years, the ability to point with high accuracy was a unique advantage with the Copernicus satellite and this has made possible a succession of valuable observations of X-ray stars in the eight years of operation.

1.4　The discovery of cosmic X-ray sources

The first clear evidence for the detection of X-rays, coming from beyond the solar system, was published in 1962, after a rocket containing Geiger counters had been successfully launched in June of that year. The objective of the experiment, conducted by the American Science and Engineering (AS & E) group led by Riccardo Giacconi, was to search for X-rays emitted by fluorescence from the Moon.*

Somewhat earlier, Herbert Friedman's group had obtained puzzling results from a rocket experiment conducted in 1956. These results were discussed at a meeting of the International Astronomical Union held in Moscow in 1958. The possibility that celestial X-rays had been detected was certainly recognized at that time, but there were also alternative explanations, such as localized production of X-rays in the atmosphere. Nevertheless, the intriguing possibility that there might be intense sources of X-rays other than the Sun acted as a catalyst in theoretical speculation about the kinds of processes that could produce the emission. It also encouraged experimenters to conceive of more sensitive types of detectors with which to make further observations. Friedman, for example, wrote in *Scientific American* in 1959 that 'Rocket astronomy has not yet undertaken the observation of celestial objects in the X-ray spectrum. Such cosmic ray sources as the Crab Nebula, have a high priority in experiments now being designed and instrumented.'

Within one year of the results obtained by Giacconi's group, Friedman and his colleagues at NRL had confirmed that there was indeed an intense source of X-rays, and that they were coming from the direction of the constellation of Scorpio. The NRL group had produced detectors with a larger collecting area and they had also introduced smaller viewing fields of the detectors (10° compared with the 100° of Giaconni's instrument). This enabled them to determine the position of any detected source more precisely. They were able to discover a second source which was in the vicinity of the Crab Nebula.

The Crab Nebula eventually became the first object outside the solar system to be identified as an X-ray source. The experiment, again by Friedman and his colleagues, was a remarkable achievement of rocket astronomy. The Nebula lies close to the ecliptic (the plane

* See *X-ray Astronomy* by R. Giacconi and H. Gursky for an account of the early literature.

of the Earth's orbit), and every ten years or so the Moon passes across the Nebula for a few lunar cycles. Although a sounding rocket remains above the Earth's atmosphere for only a few minutes, during this time the Moon moves sufficiently in its orbit for it to be used as an obscuring disc to determine the size of any X-ray emission. For the experiment to succeed, the instruments had to be pointed at the Nebula continuously and the rocket had to be launched at a pre-determined time with a margin for error of only a few seconds.

In addition to identifying the Crab Nebula as the source of the signal detected in the earlier rocket flight, the result also clearly showed that the emission was not localized in a point source. Since the distance to the Nebula is known, it was also possible to estimate the total amount of energy being radiated as X-rays. This turned out to be as much as 1,000 times the luminosity of the Sun in all wavelengths.

The Crab Nebula is a fascinating object for study in almost every part of the spectrum. It is the remains of a star which exploded in AD 1054 and became so luminous that it could be seen in the daytime sky. Chapter 6 describes how such dramatic explosions occur and what has been discovered by studying the X-ray emission, which continues for thousands of years after the explosion. We call these events supernovae, and the nebula resulting from an explosion is called a supernova remnant.

It was much more difficult to identify the strong source that had been discovered in the constellation of Scorpio. The source, which is known as Sco X-1, is never eclipsed by the Moon. Accordingly, new techniques for refining the position had to be developed. In 1964 a group of scientists from AS & E and the Massachusetts Institute of Technology (MIT) made measurements of the angular size and position of the source with a new technique called (after its inventor) the Oda collimator.* The position of the source was determined to an accuracy of about 1° and the angular size found to be less than a tenth of a degree. Unlike the Crab Nebula, there was no obvious sign that there had been a supernova explosion in the area and the nature of the X-ray source remained a puzzle.

It was not until 1966 that the mystery of the Sco X-1 source began to be revealed. In that year further refinements of the position of the source were made in another rocket experiment by the same two groups. They were able to reduce the area of uncertainty in the

* A collimator in X-ray astronomy defines a viewing direction, usually by a honeycomb or bundle of tubes in front of the detector.

source position to 1/1000 of a square degree. Astronomers at Tokyo and Mount Palomar observatories searched the area with large telescopes, and shortly after the new position became available they identified a faint blue star as the most likely candidate.

The identification of the Sco X-1 source must be regarded as a turning point in X-ray astronomy, for this star could be described as an X-ray star; its visible output is 1,000 times smaller than the energy it emits as X-rays. In comparison with the Sun the source is quite remarkable: the X-ray output is 10,000 million (10^{10}) times greater than the normal emission from the corona. It was such an unusual object that some previously unrecognized form of energy source had to be providing this intense X-ray emission.

The energy source for the Crab Nebula did not create such a problem, as it was known, for example, that the radio emission from the nebula was probably caused by high-energy particles interacting with a magnetic field. X-ray emission could also be produced by this mechanism, which is known as magnetic Bremsstrahlung or synchrotron radiation. The source of the high-energy particles was not discovered until much later, as we shall see in Chapter 6.

Optical astronomers provided the clues for the energy mechanism in the Sco X-1 source. Early observations indicated that the supposedly single star was in fact two objects in orbit about their common centre of gravity. This prompted several astronomers to suggest that the source of energy was material flowing from one of the stars to the other. For this mechanism to work efficiently, the star on which the material falls must be very much smaller but of the same, or greater, mass than a normal star. Only one type of star was known to exist in this form: a white dwarf. The gravitational potential at the surface of such a star is very high, so that material falling on to the surface gains a high velocity which is converted into heat when it reaches the surface. The temperature produced can be high enough to give radiation in the X-ray part of the spectrum.

Two other types of collapsed star had been predicted but their existence had not been confirmed. The first neutron star was discovered a year later in 1967 by Bell and Hewish who observed rapid, regular pulsations in radio sources. These were interpreted by Gold as resulting from beams of radio waves emitted in the magnetic field of a rapidly rotating neutron star. The other type of collapsed object is the black hole. We shall see in Chapter 8 that X-ray astronomy may be the only branch of the subject that can lead to the conclusive proof of the existence of these bizarre objects.

The early optical observations of Sco X-1 in fact turned out to be incorrect. The theory that the source of energy powering X-ray stars was intense gravitation around the collapsed objects fell out of fashion for a time. Later observations of other stars have convincingly demonstrated that some X-ray sources are powered by this mechanism. However, Sco X-1 remains somewhat of a mystery; although it was eventually established that two objects are in the system, it has not yet been possible to demonstrate that the X-ray source takes part in any of the regular variations characteristic of a binary star.

The identification of the Crab Nebula and the brightest source Sco X-1 provided astronomers with examples of a diffuse and a point source of X-rays, both in our Galaxy. When the first source outside our Galaxy was discovered, also by the NRL group in 1966, X-ray observations were recognized as relevant in the study of the most distant objects in the Universe. The thirty or so known sources ranged in luminosity from 10^{26} ergs/sec for the Sun to 10^{43} ergs/sec for M87, the extra-galactic source. A diffuse background of X-rays had also been discovered, which might originate from an earlier stage in the expansion of the Universe. The value of the new astronomy had been proven and, as we shall see in the following chapters, the results from the subsequent satellite experiments have been quite remarkable.

2 · The nature of X-radiation

2.1 Historical introduction

While it is now clear that X-rays are a form of radiant energy called electromagnetic radiation and as such belong to the same family as visible light, radio waves and infrared radiation, this relationship was not understood at the time of their discovery. X-rays were first identified by Röntgen on November 8 1895 at Würzburg in Germany. The penetrating power of the radiation was surprising and the potential for use in medical radiography was quickly recognized and exploited. Although the rays did not respond like charged particles to electric and magnetic fields, it was also recognized that lenses and prisms did not cause X-ray beams to deviate in the manner of visible light although they were detectable by photographic film. Thus the sum of the available evidence was such that even shortly after the discovery of X-rays Röntgen was able to propose that some sort of wave phenomenon was involved.

The wave nature of X-radiation was firmly established by the work of Von Laüe in 1912 and the Braggs in 1913. They showed that X-rays were diffracted by atoms in crystal lattices. This sort of interaction, which we will discuss briefly later, can happen only if transverse waves are involved. In addition, the Bragg law—which relates the wavelength of the diffracted rays to the angle the ray path makes with the crystal—allowed Moseley and others to investigate the nature of both X-rays and the diffracting crystals in considerable detail. One of the great triumphs of the X-ray-diffraction technique was the analysis of the DNA molecular structure in the 1950s by Watson and Crick—work for which they were awarded a Nobel prize. Many other significant applications of X-ray physics were established in the decades following Röntgen's discovery. These yielded a variety of studies in atomic, nuclear and solid-state physics. Intensive work in these and other areas has led to the development of many sophisticated techniques for the detection and analysis of X-radiation. Thus, while X-ray astronomy—as we have seen—had to wait for the dawn of the space age, the many other areas of application of X-ray techniques ensured that when the time came, these techniques would be available for rapid deployment in the study of the Universe.

In this chapter we will briefly discuss the nature of X-radiation. Because X-rays differ significantly from visible, infrared, and radio radiation, it will be necessary to describe some of their unique properties so that their impact on astronomy may be more easily appreciated. After a discussion of X-ray properties, we will describe the interactions of X-radiation with matter and, in particular, with the atmosphere of the Earth. X-rays require rather special techniques for their detection and these will be mentioned briefly. If certain stringent conditions are fulfilled it is possible to reflect X-radiation from highly polished metal surfaces although this does involve considerable difficulties. For this reason the nature of X-ray telescopes, their development and their use by X-ray astronomers will be described.

2.2   X-rays and their place in the electromagnetic spectrum

When the wave nature of X-rays was established, it became clear that they were a form of electromagnetic radiation similar in kind to radio, infrared, visible, ultraviolet and gamma radiation. The location of the X-ray band in this range or spectrum of radiation is indicated in Fig. 2.1. All electromagnetic waves travel in a vacuum at a velocity (c) of 299,793 km/sec, which is known as the velocity of light. Electromagnetic waves are in fact a form of energy carried through space by varying electric and magnetic fields. These variations are systematic in direction and amplitude. They are wave-like and the properties of the different electromagnetic radiations depend very largely on their wavelengths (λ) or frequency (υ). For all such radiation, the relation $c = \upsilon\lambda$, or velocity = frequency times wavelength, holds. Thus gamma and X-rays have the shortest wavelengths and the highest frequencies (see Fig. 2.1) while, conversely, radio waves have long wavelengths and low frequencies.

Although electromagnetic radiation is a wave phenomenon, the quantum theory tells us that energy is transferred by the radiation in packets or quanta and not continuously. According to Planck's quantum theory, the energy of these quanta is given by the relation

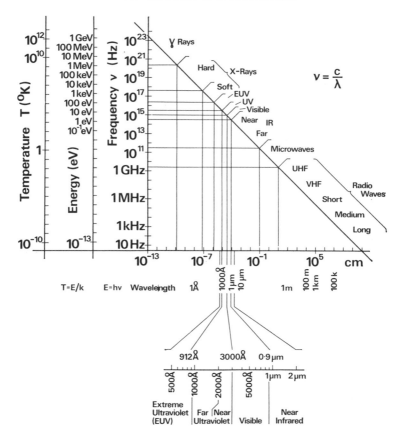

Fig. 2.1 The electromagnetic spectrum. (From an original by Lucio Chiapetti of Milan University, and D. Hoyle, Mullard Space Science Laboratory, University College, London)

Energy (E) = Constant (h) times frequency (υ) or $E = h\upsilon$, where the quantity h is known as Planck's constant. Fig. 2.1 indicates the units of wavelength, frequency and quantum energy that are used throughout the electromagnetic spectrum. For those spectral regions where the quantum or photon* energy is small (i.e. less than one electron volt (1 eV)), individual photons are not detectable and departures from the wave theory are barely noticeable. At high photon energies (it is clear from Fig. 2.1 that X-rays are in this category) the quantum theory becomes essential to our understanding of the phenomena and electronic detectors in particular register

* Quanta of electromagnetic radiation are also called photons.

X-radiation by counting single photons. However, it remains important to bear in mind the dual (wave and quantum) nature of electromagnetic radiation and hence Fig. 2.1 includes wavelength,
frequency and photon-energy units.

The spectral range of principal interest in X-ray astronomy is
indicated in Fig. 2.1. In photon-energy terms, the X-rays of
interest to astronomy are of a rather lower energy (are somewhat
'softer') than those used in medical and industrial applications. This
is because the luminosity (number of X-ray photons emitted per
second) of celestial X-ray sources decreases rapidly with increasing
photon energy. Although 'hard' X-ray and gamma ray astronomical
observations are carried out, the relatively low source intensities
make these disciplines difficult to pursue. The low photon-energy
limit to the X-ray range is imposed, as we shall see later, by the
absorption of X-rays in the gas and dust of our Galaxy.

2.3 The interaction of X-rays with matter

Before discussing the production of X-rays in astronomical sources
(see Chapter 3), it is important to examine the ways in which X-
radiation can be absorbed by matter. The absorption of the radiation
both in the Earth's atmosphere and in the gas of the Galaxy has a
highly significant impact on the subject. Furthermore, it is necessary
to absorb X-rays in matter in order to detect them.

If a beam of X-rays, of intensity I_0, is directed on to a slab of
matter of thickness x, then each of the η atoms in the slab has an
effective area or 'cross-section' σ for absorbing X-rays and the
situation will be as depicted in Fig. 2.2(a). The area $\eta\sigma$ in the slab
removes X-rays from the incoming beam and thus the outgoing
intensity is $I_0 - \delta I$, that is the beam intensity has been reduced by
a small amount δI.*

Three different physical processes contribute to X-ray absorption
in matter and the operation of these is illustrated schematically in
Figs 2.2(b), (c) and (d). The relative importance of the three processes
depends on the photon energy and the atomic number of the
matter atoms involved. In the *photo-electric* effect, the energy of the
incoming photons is used to remove bound electrons from the atoms
in the slab of material (see Fig. 2.2(b)). The value of σ, the absorption cross-section, decreases with photon energy for a given absorber
but as the atomic number of the element increases so too does the

* It can be shown that $I = I_0 \exp(-\eta\sigma)$ is the emerging beam intensity.

value of the cross-section for a fixed photon energy. Thus, light, low-atomic-number elements like hydrogen and carbon are much less effective absorbers than the heavy elements. It is important to remember that it is the total number of atoms in the path of the X-ray beam that determines the overall reduction in intensity. While a gas has many fewer atoms per unit volume than a solid, if the radiation traverses a long path in a gas it can still be absorbed quite effectively.

When a photon interacts with an atom by the photo-electric effect, the result is the liberation of an electron from the target atom.

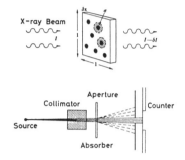

a) Absorption and Scattering of X-rays   b) Photoelectric Effect

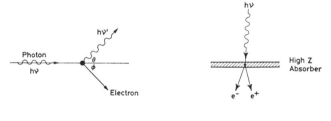

c) Compton Effect   d) Pair Production

Fig. 2.2 X-ray interactions with matter. (a) A beam of X-rays of intensity I strikes an absorber of unit area and thickness δ. An effective area, σ, is presented to the beam by each atom and the emerging beam has intensity $I-\delta I$. In addition to absorption, scattering of radiation is also depicted. (b) An incoming X-ray photon is shown removing a bound electron from an atom. (c) A photon gives up energy to an electron and leaves the interaction site at an angle θ to its original direction. The electron leaves the site at an angle φ to the photon direction. (d) A high-energy gamma-ray photon undergoes pair production interaction in an absorber of high atomic number (Z)

This electron acquires and carries off the energy of the photon. This offers the possibility of detecting the incoming X-ray photons. Most detectors used in X-ray astronomy operate in this fashion. A number of examples are discussed below.

Another photon–matter interaction is called the *Compton effect* (see Fig. 2.2(c)). This interaction becomes significant at higher photon energies (i.e. $E\upsilon > 50$ KeV). In these circumstances, the photon energy is much greater than the energy that binds the electrons into atoms and so the target electrons may be thought of as virtually free electrons. The incoming photon gives up some of its energy to the electron and leaves the interaction site at some angle θ to its original direction. The recoiling electron leaves in a direction that makes an angle φ with the original photon path and carries away momentum equal to the momentum difference between the original and outgoing photon. For a given element, the Compton effect takes over increasingly from the photo-electric effect as the photon energy increases. It is important to notice that this process does not totally stop the incoming photons but redirects or scatters them. Hence, while the photon remains in existence with a lower energy its direction is changed. However the total cross-section for the interaction gives a valid measure of the likelihood of an X-ray photon being removed from the beam direction.

A final point worth noting concerns polarized radiation. If radiation is polarized, the directions in which the electric and magnetic fields can vary are fixed in space. In this situation, there is a maximum probability that the scattered photon and recoil electron are ejected approximately at right angles to the direction of the electric vector of the incoming X-rays.

At even higher photon-energies ($E\upsilon > 1$ MeV) it is possible for a photon to interact in matter so that it disappears and is replaced by an electron–positron pair. This process is known as *pair production* and its occurrence is illustrated schematically in Fig. 2.2(d). Pair production takes place in the region of electric field around a nucleus. The cross-section increases with the square of the atomic number (Z) of the material in which the interaction takes place. This is because the size of the region of strong electric field around the nucleus also increases with Z^2. A roughly equal division of photon energy between the electron and the positron is the most likely outcome.

Following their creation, the electron and positron may come to rest in the absorber. The positron will then interact with another electron so that both are annihilated and a photon is created. Because of the dependence of the cross-section on Z^2, absorbers of high atomic number (e.g. lead) are usually employed. In fact the pair-production process is mainly of interest in gamma-ray astronomy, since it can occur only if the incoming photons have energy greater than 1 MeV. For this reason, it will not be of great importance in the present book.

One further process that we will encounter, namely *Thomson scattering*, involves the redirection of a beam of incoming X-rays by free electrons. The electric vector of the X-radiation field forces the electrons to oscillate. These electrons then re-emit in accordance with the classical laws for radiation by moving electric charges. No energy is lost to the electron in this process; it is a pure scattering interaction. This process is important in the construction of X-ray polarimeters since the cross-section becomes a maximum when the outgoing photon direction makes an angle of 90° with the electric vector of the incoming wave. Thomson scattering is also of importance to the propagation of X-rays through the dense atmospheres of certain astronomical X-ray sources, which will be discussed later.

We will encounter many applications of the ideas about absorption and scattering in later chapters. One which we will consider now is the absorption of X-rays by the atmosphere of the Earth. The fact that X-rays (along with ultraviolet and gamma radiation) are absorbed before they can reach the Earth's surface, has forced the development of these branches of astronomy to wait for the availability of space vehicles such as balloons, sounding rockets and Earth satellites, which can carry detection equipment above the absorbing layers of the atmosphere. The need for these vehicles is shown clearly in Fig. 1.1 where the height above the Earth's surface at which the incoming radiation is absorbed is shown plotted against photon energy and wavelength. The heights at which balloons, sounding rockets and satellites usually operate are discussed below.

Balloons can remain at heights of around 30 km for periods of about forty-eight hours, although developments now in progress could extend this interval considerably by using super-pressure balloons for around-the-world flights. From the standpoint of X-ray astronomy, a balloon-borne instrument can begin to detect incoming radiation only at energies above 20–25 KeV, whereas, as we shall see, most of the important results obtained so far have required observations in the 1–10 KeV energy range.

A variety of vertical sounding rockets (as illustrated in Plates

3 and 4) have been used by X-ray astronomers during the past three decades to study X-rays from the Sun and other cosmic sources. They include the V-2, captured examples of which were used, as we have seen, by the Americans after the Second World War to make, among other observations, the first detection of solar X-rays. Sounding rockets can reach an adequate height (i.e. greater than 120 km) to carry out X-ray observations but have the significant disadvantage that they can remain at astronomically useful altitudes for, at most, four or five minutes. Nevertheless, X-ray astronomy owes its very existence to observations made with rocket-borne instrumentation.

Orbiting satellites have enormous advantages in that they remain at their operational altitudes for periods of up to several years. Although they have been generally limited in volume and weight compared with balloons and sounding rockets, their ability to observe for long periods has more than compensated for their other limitations. Three examples of satellites devoted to X-ray astronomy are the American Uhuru and OSO and the British Ariel V spacecraft. Their role in the subject has been of crucial importance. The Ariel V satellite is shown in Plate 5.

2.4   X-ray detectors and spectrometers

Although medical X-ray systems produce photons of 50–100 KeV, which can penetrate human tissue, the energy region of greatest interest in X-ray astronomy is 1–10 KeV. We have seen that these softer X-rays are relatively easily absorbed in matter, a fact which makes them difficult to detect. It is also true that X-rays cannot be handled by lenses and only reflected if very stringent conditions are met (see Section 2.2). It is therefore impossible to undertake X-ray astronomical observations with telescopes similar to those used in optical and radio astronomy. Both for this reason, and because of the need to place instruments above the Earth's atmosphere, the development of X-ray astronomy has proceeded in a very different manner from that of its optical or radio counterparts.

The most commonly used detector in X-ray astronomy has been the gas-filled proportional counter (see Fig. 2.3). This device admits X-rays through a large ($\sim$ 2,000 cm^2) thin metallic or plastic window made of low-atomic-number (Z) material, so that its photoelectric absorption cross-section is a minimum. It is then filled with a high-atomic-number gas so as to increase the chance of absorbing and detecting X-rays within the detector volume. The photo-electric absorption of a photon anywhere in the detector is followed by the release of a photo-electron. This electron has sufficient energy to collide with other atoms and remove additional electrons from them. Thus the absorption of a 1 keV photon in the gas will result in the production of about thirty electrons. It is important to note that this number is *proportional* to photon energy. Thus a 2 keV photon will liberate sixty electrons. A high positive voltage ($\sim$ 2,000 V) is connected to the anode, a tungsten wire of 50 micrometres diameter.

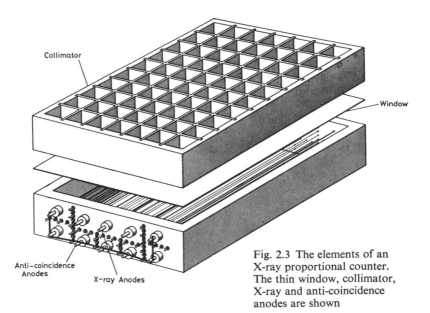

Collimator

Window

Anti-coincidence Anodes

X-ray Anodes

Fig. 2.3 The elements of an X-ray proportional counter. The thin window, collimator, X-ray and anti-coincidence anodes are shown

The field generated by this voltage causes the initially created electrons to move rapidly towards the anode. On their way they collide with atoms and liberate further electrons, which are in turn accelerated by the field and liberate even more electrons from atoms; this leads to the formation of a charge avalanche containing as many as 300,000 electrons. The amplitude of this charge impulse, which is formed in about 10^{-6} sec, can be measured electronically, and thus in addition to registering the presence of an X-ray photon the proportional counter can also measure the photon energy. In practice, statistical fluctuations in the size of the charge avalanche cause an uncertainty of around 20 per cent in a measurement of the energy of a 1 keV photon although this decreases with energy to about 7 per cent at 10 keV. So the proportional counter cannot compare in quality of wavelength or energy resolution with systems used in the ultraviolet and visible spectral ranges. Nevertheless, a great deal of our present understanding of the nature of X-ray sources is based on data obtained with proportional counter detectors.

The main purpose of the proportional counter is to detect X-ray photons which enter through its window. However, above the Earth's atmosphere, there are substantial fluxes of cosmic-ray-charged particles and the detector can respond to these in a manner that is frequently indistinguishable from the way in which it responds to X-rays. Cosmic-ray-induced events in the detector can be recognized and eliminated by identifying the occurrence of simultaneous events in both the main and guard cells of the detector (see Fig. 2.3). Since cosmic-ray particles have much higher energies than the X-ray photons, they are likely to travel through both a main and a guard cell, be detected in both and so be eliminated by this double signature. Cosmic-ray events may also be differentiated from X-ray events on the basis of much longer pulse-rise time. A good background reduction system can eliminate as much as 98 per cent of the particle background events registered in the detector.

At energies above 30 keV it becomes increasingly difficult to achieve adequate stopping power with gas-filled detectors. In this situation, a *scintillation* detector is usually employed. This comprises a sodium iodide or caesium iodide crystal, which responds to an incoming X-ray by generating flashes of light that can be detected by a photomultiplier tube placed in contact with the crystal. The intensity of each light pulse is proportional to photon energy. Silicon solid-state detectors are also used in X-ray astronomy. They are basically reverse-biassed junction diodes. Photons that enter the depletion region create hole–electron pairs, which can be registered as a charge pulse. In order to achieve good photon-energy resolution (up to three times better than that of the proportional counter) detector size must be kept small. Therefore, these detectors can only be used effectively with X-ray telescopes (see Section 2.2).

In view of the difficulties involved in constructing X-ray telescopes, up to now observations have been undertaken mainly using detectors fitted with simple mechanical collimators of the kind illustrated in Fig. 2.3. These collimators are made of thin parallel metal plates, arranged in the manner shown to give rectangular fields of view of typically 1° by 10° on the sky. We will see, in later chapters, examples of how detectors equipped with simple collimators respond to X-ray sources.

In addition to simple collimators, a variety of modulation collimators have been used. Made with fine-mesh grids, they permit source positions to be measured to ten arc second accuracy. Among devices of this kind, the rotation modulation collimator has been particularly successful in X-ray astronomy. By rotating the collimator above the detector, the photons received from a source are modulated in a manner that depends on the angle between the source position and the axis of the collimator. The frequencies of modulation can be extracted in subsequent data analysis and the source position measured.

We have stated that the energy resolution of the proportional counter is poor, particularly at low photon energies. It will become clear later that some X-ray sources emit spectral lines or groups of photons all having essentially the same energy. These lines may also appear as absorption features in the radiation output of a source. While the proportional counter represents these features as broad, and spread over a significant energy range, it is possible to obtain much higher energy resolution by means of the Bragg crystal spectrometer. If the planes of atoms in a crystal are separated by a distance d, then radiation of wavelength will be reflected from the crystal surface when the angle of grazing incidence (the Bragg angle θ) equals the angle of grazing reflection and when the Bragg law $2d \sin \theta = n\lambda$, ($n$ is an integer) is obeyed. Although Bragg spectrometers are rather inefficient at reflecting photons, they represent virtually the only way that it is possible to achieve high-energy resolution. We will describe the results obtained in solar studies with Bragg spectrometers in Chapter 3.

2.5 The reflection of X-rays from metal surfaces

While X-ray astronomy has made considerable progress through the use of mechanically collimated proportional counters, the performance of these systems is extremely limited in comparison to that of an optical telescope equipped with a good spectrometer. However it is possible to reflect X-rays from polished metal surfaces. The mechanism involved is illustrated in Fig. 2.4. A glass–air interface has a refractive index $\mu = 1.567$ for visible light. From the law of refraction, namely $\sin i / \sin r = \mu$, it is possible to show that when the angle i is greater than 42°, the light is reflected back internally into the glass block. This effect can also occur with X-rays but, for a vacuum–metal interface, the appropriate value of μ is 0 9994—that is less than 1 at very nearly unity. In this situation total *external* reflection can occur when the angle between the incoming ray path and the metal, the critical angle of grazing incidence (i_g) for such reflection, is less than about 2° (see Fig. 2.4(d)).

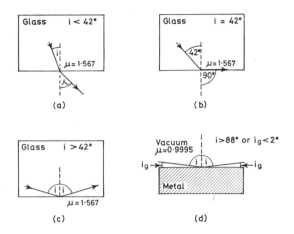

Fig. 2.4 X-ray reflection visible light is shown: (a) leaving a glass block; (b) internally reflected along the surface of the block; (c) reflected back into the block; (d) reflected from a metal surface when the rays are at grazing incidence

The actual behaviour of reflection at a metal surface shows two extreme forms, which are illustrated in Fig. 2.5, in which the reflectivity of two different surfaces is plotted against the angle of grazing incidence for a fixed X-ray wavelength. For the material of lower atomic number (Z), the reflecting power is high for small glancing angles but then drops rapidly above a value of glancing angle called the critical angle (θ_c). For the higher Z material, the behaviour of reflection with angle is much less sharp and the reflecting power gradually declines over a larger glancing-angle range. These graphs indicate the severity of the limitations that confront the designers of X-ray-reflecting optical systems.

In 1952, before the beginning of X-ray astronomy, the German scientist Hans Wolter proposed a number of X-ray-reflection systems that might be used to implement an X-ray microscope.

These have since been adapted for use in X-ray astronomy. The simplest telescope configuration is a single paraboloid of revolution whose use in X-ray astronomy was first proposed by Giacconi and Rossi in the United States in 1960. This reflector, which is illustrated schematically in Fig. 2.6(a), consists of a deep paraboloid of revolution. X-rays parallel to the axis of the reflector are brought to a focus at F but rays making an angle α with the axis cannot form an image because of severe aberrations. Instead they are concentrated

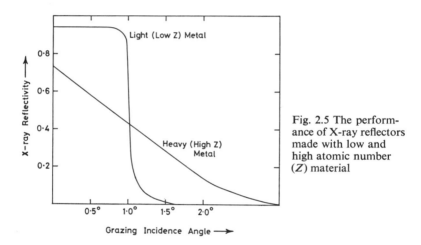

Fig. 2.5 The performance of X-ray reflectors made with low and high atomic number (Z) material

into an annulus in the focal plane whose outer radius is $L\alpha$ and whose inner radius is $L\alpha/2$. If a detector with a window of radius $L\alpha$ is placed at the focus, then only rays that make an angle of less than α with the axis will be detected and so a field of view can be defined on the sky. (Values of α of as little as 30 arc seconds were used in the X-ray telescopes that were flown on the Copernicus satellite, see Plate 2.) Although the small field of view compares very favourably with that achievable by mechanical collimators, the need to maintain the angle of grazing incidence below 2° means that only a small fraction of the aperture (diameter D, Fig. 2.6(a)) provides useful collecting area.

Because of the severe aberrations associated with a single grazing-incidence reflection, it is not possible to form an X-ray image using a simple paraboloid. Wolter described two image-forming systems which involve two reflections. The first of these is illustrated schematically in Fig. 2.6(b). Known as the Wolter type-I system, it uses two reflections from a paraboloid and a hyperboloid of revolution. Incident photons are first reflected from the paraboloid, then from the hyperboloid and are finally brought to a focus. Although the two reflections greatly reduce the effect of aberrations, they do not eliminate them entirely. However, on axis, angular resolutions of around 1 arc second can be achieved. The largest X-ray telescope so far made was launched late in 1978 on the second High Energy Astronomy Observatory (Einstein HEAO-2) by Giacconi and his colleagues (see Plate 6). It employs a nest of four Wolter type-I telescopes mounted one inside the other to increase the collecting area and it has an angular resolution of a few arc seconds. However, imaging telescopes have already been used with great success in studies of the Sun's X-ray emission. An example of a solar X-ray image obtained by Krieger and his colleagues at AS & E is shown in Plate 7 (see Chapter 3). In this work an angular resolution of about 1 arc second has been achieved.

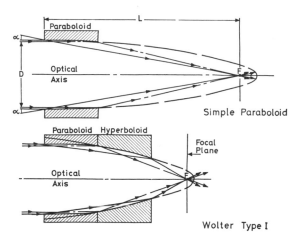

Fig. 2.6 Optical design of X-ray telescopes.
(a) The simple paraboloid. (b) The paraboloid-hyperboloid combination

It is, however, very difficult to achieve arc second performance at X-ray wavelengths. Working at wavelengths in the range 4,000–8,000 Å the best optical telescopes require that, for arc second performance, the maximum deviation from a perfectly figured and polished (i.e. perfectly smooth) surface should be less than 200 Å. We have seen that the X-ray wavelengths of greatest interest lie in the range 1–20 Å. If the smoothness requirement is scaled proportionately with wavelength, the necessary maximum deviation would be less than 0.5 Å or 5×10^{-5} micrometres! Fortunately, it is the projected deviation in the direction of the incoming radiation that is of importance. Since the radiation must be at grazing incidence in order to be reflected, the deviation requirement may be relaxed by a factor equal to the sine of the glancing angle, which sets the maximum permitted deviation at around 25 Å or 0.003 μm. Considerable work has nevertheless been necessary on figuring and polishing techniques and on choice of materials in order to obtain X-ray telescopes of arc second resolution.

Given the existence of imaging X-ray telescopes, it is necessary to register the images formed and transmit them from the space vehicle back to Earth. Photographic film has excellent spatial resolution and a large information capacity. It has been used to good advantage with rocket-borne X-ray telescopes and on the Skylab mission (see Chapter 3) to obtain high-quality X-ray images of the Sun. However, it has a low detection efficiency and in addition is affected by the charged-particle background in a manner that does not permit the use of any background reduction technique. It is also not practical to recover film from a spacecraft that is in orbit for several years. For these reasons its use is likely to be confined to the registration of solar data on short-duration missions.

Proportional counter detectors can be adapted to measure the position of arriving photons in both one and two dimensions. Such a detector developed in our laboratory is illustrated schematically in Fig. 2.7(a). A strong electric field between the grid and the resistive plate leads to the creation of a charge avalanche in response to the detection of an incoming X-ray photon. Electronic amplifying circuits are connected to four points on the resistive disc and by observing the different time development of the signal in each circuit, it is possible to measure the original arrival position of the charge on the disc with an accuracy of better than 0.2 mm in two directions.

Even better spatial resolution can be achieved with a device called a *microchannel plate*. Single-channel multipliers (see Fig. 2.6(b)) have been used as X-ray detectors for some time. X-ray photons, which strike the tube walls at one end, liberate electrons. These electrons move along the tube gaining energy from the electric field and colliding with the tube walls. At each collision further electrons are liberated so that eventually a large secondary-electron charge pulse is collected at the far end of the tube. In a channel plate a very large number of single-channel multipliers are stacked together. Bundles of channels with tube centre-to-centre spacings of 15 micrometres are available (see Fig. 2.7(c)). In order to obtain a stable high-gain operation, it is necessary to use a pair of plates with one mounted at a slight angle to the other. The resulting

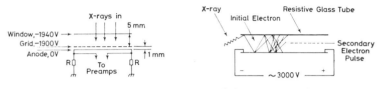

a) Parallel-Plate Proportional Counter (schematic)

b) Single-Channel Multiplier

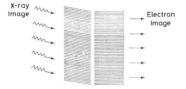

c) Chevron Microchannel Plate

Fig. 2.7 Operation of the imaging proportional counter and microchannel plate detectors. (a) Schematic diagram of the imaging proportional counter. (b) The operation of a single-channel electron multiplier. (c) Combination of a number of small single channels to form a microchannel plate

electron image can be read out, either with a resistive disc as described above for the proportional counter, or with an array of closely spaced wires. Both of the electronic image-forming detectors have the advantage that the X-ray event positions can be electrically encoded and transmitted over the spacecraft radio link to Earth. Recovery is not required, as would be the case with photographic film.

3 · X-rays from the Sun

3.1 The Sun and its atmosphere

The Sun is much more than our nearest star. It sustains life on our planet and therefore played a prominent role in the religions of many early civilizations. In the twentieth century its scientific study has assumed considerable importance and it has thus become a natural target for X-ray observations.

The Sun is a very ordinary star; there are around 10^{10} similar stars in our Galaxy. It was formed from a collapsing cloud of gas about 3×10^9 years ago and has now reached a relatively stable state of existence, in which it has halted further collapse by obtaining energy from the nuclear fusion of hydrogen in its core. We will discuss the evolution of stars in greater detail in Chapter 6. The core temperature is high enough ($\sim 15 \times 10^6$ K) to permit these nuclear reactions but falls rapidly nearer the surface and at the surface of the Sun it drops to around 5,700 K. Because of the nuclear reactions that go on there, the core should also emit a large flux of neutrinos, esoteric elementary particles which have no mass but which can possess momentum and energy. These particles are almost, but not quite, impossible to detect. Davis and co-workers, who have installed a large detection system in a deep mine in the west of the United States, have been working on this problem for many years. Their results have important implications for our understanding of the Sun's structure. Energy is transferred from the core first by radiation, but near the surface the transfer mode changes to convection, so in addition to the radiation from the Sun's surface—or photosphere—there is a good deal of mechanical energy crossing the outer surface. The importance of this will become clear later.

Seen in visible light, the Sun is a spherical body which subtends an angle of 32 arc minutes at the Earth. Its diameter is 1.4×10^6 km. A number of solar parameters, with their values, are listed in Table

3.1. While the photosphere is easily detected and studied, in addition there exists a tenuous outer atmosphere of low-density gas, which extends for several solar radii above the surface. The first layer of this atmosphere is only a few thousand kilometres thick and is called the *chromosphere* or colour sphere because of its bright red colour which may be seen briefly, at times of total eclipse by the Moon. The outer layer of the atmosphere is called the corona (see Plate 1). Once again it is visible only during total eclipses or with the aid of special instruments called coronographs that are equipped with occulting discs to mask the photospheric light. It is this outermost region of the solar atmosphere with which we will be mainly concerned. The visible-light output of the corona is more than 1,000,000 times less than that of the photosphere and hence studies of this region at visible wavelengths are fraught with considerable difficulty.

At first sight it would seem surprising that a relatively ordinary star like the Sun with a surface temperature of 5,700 K should be able to produce X-rays. We shall see later that gas temperatures in the range above 10^6 K are required for X-ray generation. However, even before the possibility of observations from space there was evidence which strongly suggested that the corona had an

Table 3.1   Some properties of the Sun

Radius $(R_\odot)$	$=$	7.0×10^8 m $(7 \times 10^5$ km$)$
Volume $(V_\odot)$	$=$	1.4×10^{27} m^3
Mass $(M_\odot)$	$=$	2.0×10^{30} kg
Mean density $(\rho_\odot)$	$=$	1.4×10^{-9} kg m^{-3} $(1.4$ gm cm$^{-3})$
Radiation output $(L_\odot)$	$=$	3.8×10^{26} W
Surface escape velocity	$=$	6.18×10^5 ms^{-1} $(618$ km s$^{-1})$
General magnetic field	$=$	1–2×10^{-4} T $(1$–2 gauss$)$
Mean distance from Earth	$=$	1.5×10^{11} m $(1.5 \times 10^8$ Km$)$
	$=$	1 astronomical unit (AU)
	$=$	$93,000,000$ miles
Angular diameter at mean distance	$=$	32 arc minutes
Angular scale	$=$	7.25×10^5 m $(725$ km$)$ for 1 arc second
Effective temperature	$=$	5770 K
Age	$=$	$\sim 5 \times 10^9$ years

unusually high temperature. Certain *absorption* lines from the photosphere called the Fraunhofer lines appeared again, but very much broadened, in the coronal radiation. It was believed that the broadening was due to scattering of the photons by high-velocity (i.e. high-temperature) electrons in the corona.

This observation was supported by the work of Edlén who, in 1942, solved a major outstanding puzzle regarding the nature of the corona. Studies of the coronal visible radiation had revealed certain *emission* lines that defied identification. Usually, when sharp line features are seen in the radiation output of a source, it is possible to say with certainty that each line was produced by a re-arrangement of the electrons in a specific element. These particular lines did not correspond to those of any previously known element and so at first it was thought that a new element—it was named Coronium— had been discovered in the Sun. However, Edlén demonstrated that a new element was not present, by identifying the lines as being produced in highly ionized iron ions. When an atom is singly ionized, one of its bound electrons has been removed, usually due to a collision with an energetic free electron. If doubly ionized, two electrons have been removed, if trebly ionized three have been removed and so on. Astronomers use Roman notation to denote stages of ionization; thus Fe I is neutral iron, Fe II is singly ionized iron, Fe III is doubly ionized. The emission lines which Edlén identified came from Fe X (nine electrons removed) and Fe XIV (thirteen electrons removed). Iron has a total of twenty-six bound electrons and in order to remove half of them, gas temperatures in the range 1–2 $\times$ 10^6 K are required. Thus Edlén's discovery provided the strongest evidence yet that the solar corona consisted of high-temperature gas. As we shall see below, such a high-temperature gas is a copious emitter of X-radiation.

Before describing the production of the radiation, it is worth discussing the meaning of temperature in the solar corona and the nature of the hot gas. The temperatures encountered on the Earth's surface are low and matter is not ionized; atoms have their full complement of bound electrons. But at a temperature of 10^6 K, matter is entirely gaseous and the most abundant element, hydrogen, exists only as protons and electrons; all hydrogen atoms have been ionized. A gas of this sort, which is composed of ions and electrons while remaining electrically neutral overall, is known as a plasma. Temperature is a convenient quantity by which to characterize the energy of the particles. In a steady state the electrons and ions will have velocities given statistically by the Maxwellian velocity distribution. The higher the temperature, the higher the particle velocities and hence the greater the energy in the plasma.

Although virtually all the hydrogen in the corona is ionized, the ionization stages of the other elements are strongly dependent on the electron energy and therefore on the temperature. Bound electrons are removed from atoms or ions by collisions with energetic free electrons. The higher the free-electron energy, the greater the number of bound electrons that can be removed from a given element. For example, oxygen atoms normally have eight bound electrons. At a gas temperature of 10,000 K, oxygen will still be mainly in its neutral atomic form. At 100,000 K the most common oxygen stage will have three electrons removed and will be written O IV, in accordance with the notation defined above. At 10^6 K, six electrons will have been removed and the notation is O VII or helium-like oxygen. At 3.10^6 K the predominant species is O VIII or hydrogen-like oxygen. Finally, at temperatures above 10^7 K, all of the electrons will have been removed. All the elements go through similar progressions with increasing plasma temperature. Recall, for example, Edlén's discovery mentioned above, which demonstrated that at coronal temperatures, iron exists as Fe X and Fe XIV. But these ions emitted lines in the visible part of the spectrum. We will see below that the lines emitted by most of the coronal ions are at X-ray wavelengths and that X-ray observations can tell us a great deal about the nature of the Sun's outer atmosphere.

3.2 The solar corona and its X-ray emission

Given the existence of high-temperature plasma in the corona, it was clear that such a gas should emit X-rays in the 1–10 keV photon energy range. The United States Naval Research Laboratory (NRL), under the leadership of E. O. Hulburt, began the search for solar X-rays in 1946 using V-2 rockets to carry their detectors 100 km above the Earth's atmosphere (see Fig. 1.1). The first positive detection of X-rays was obtained by Burnight on 6 August 1948. He employed a piece of photographic film placed behind a thin beryllium foil. The beryllium kept out visible and ultraviolet radiation but admitted the solar X-rays, which were detected by the film. The NRL remained the only laboratory working in this field of research for almost a decade. Their experiments quickly established that the Sun was indeed a powerful source of X-rays, the intensity of which varied with the solar-activity cycle. Before describing the nature of the Sun further, we will discuss the mechanisms for the production of X-rays in a high-temperature plasma.

As pointed out in the previous section, a plasma is composed of free electrons and ions. These interact with each other to produce X-rays in a variety of different ways. If an electron approaches close to an ion, it is decelerated in the ion's electric field and will emit radiation. An electron with energy of several keV will emit a photon of comparable energy and an assembly of electrons and ions will emit a continuous spectrum of radiation similar to that shown in Fig. 3.1. This continuous emission is called *Bremsstrahlung* or braking radiation (from the German). As the plasma temperature is increased, the spectrum falls off less steeply with X-ray energy. In addition to Bremsstrahlung, there is another quasi-continuous emission process, known as the *free-bound* process. Here a free electron approaches an ion, but does not escape after the interaction and is instead captured into a vacant quantum shell surrounding the nucleus. The shape of the spectrum is similar to that produced by the Bremsstrahlung process but is interrupted by features called *edges*. The energies at which these edges occur correspond to those of the atomic shells into which the incoming electron has been captured. Examples are shown in Fig. 3.1.

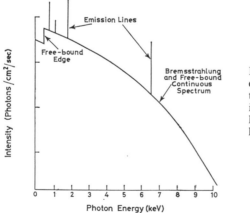

Fig. 3.1 The X-ray emission from a high-temperature gas including Bremsstrahlung, free-bound and line radiation

The final process is that of *line emission*, about which we have already spoken. Here an incoming electron is neither decelerated nor captured, and it does not remove an already bound electron.

Instead it rearranges (or excites) a bound electron to a new level (or shell) within an ion. After a very brief interval, the excited bound electron falls back to its old (or ground) state with the emission of a photon whose energy corresponds to the interval between the ground and the excited level. A number of such photons emitted

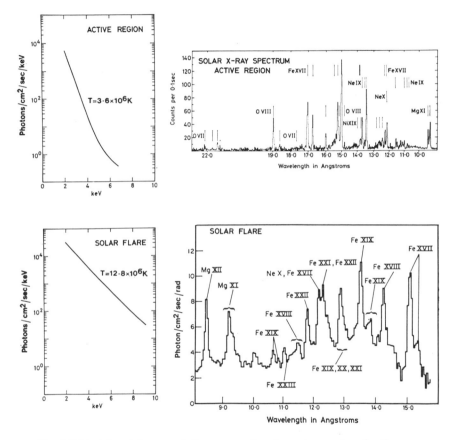

Fig. 3.2 (a) *(above)* Low-resolution proportional counter spectra of active-region material and the higher temperature gas created in a solar flare. X-ray intensity is plotted against photon energy for gas at two different temperatures. Data obtained by the Lockheed Palo Alto Laboratory's instrument on the OSO–8 satellite. The spectra appear continuous due to the low instrument resolution. (b) *(below)* High-resolution spectra of active region and flare gas obtained with Bragg crystal spectrometers. Intensity is plotted against wavelength. Individual emission lines can be discerned. The active region spectrum was obtained by Parkinson and co-workers during a Skylark rocket flight. The flare spectrum was obtained by the US Naval Research Laboratory's instrument on OSO–6

by the plasma and all having the same energy constitute a spectral line. Examples of lines are also shown schematically in Fig. 3.1. Each line corresponds to a particular electron rearrangement or transition in a specific atom or ion. The intensity of a transition (i.e. the number of photons per second) depends strongly on the plasma temperature—recall the comments regarding ions in the previous section in which we pointed out that as the temperature increases, the ionization state of the plasma moves to successively higher stages of ionization for all elements. Hence the intensity of a transition is doubly influenced by plasma temperature, which also determines whether or not the required ionization stage exists.

Thus, a high-temperature plasma will emit X-rays due to three different types of interaction between electrons and ions. Further, the nature of the spectrum is set by the ion species that are present and so the shape and content of a spectrum depends strongly on the plasma temperature as well as on the elements present. To illustrate the discussion we will now consider some observed solar X-ray spectra which are shown in Figs 3.2(a) and 3.2(b). Unfortunately a difficulty becomes apparent; namely, that our ability to observe a spectrum depends on the energy resolution and sensitivity of the detection system that is used. Fig. 3.2(a) shows spectra obtained with a proportional counter detector, a device of good sensitivity but poor energy resolution which was described in Chapter 2. The two spectra shown are characteristic of high- and low-temperature plasma. But although there are many lines present, particularly at lower photon energies, individual line features cannot be resolved. However, the change in slope of the spectrum with increasing temperature is apparent. These spectra were obtained by Loren Acton and his colleagues using the Lockheed proportional counter spectrometer on the OSO-8 satellite. Although individual lines cannot be resolved, it is possible to make a rough estimate of the plasma temperature by considering regions of the spectrum that are free of strong lines.

Much more information can be obtained from high-resolution Bragg crystal spectrometers. These instruments have adequate performance to resolve individual lines, as is demonstrated in Fig. 3.2(b). The first spectrum, obtained by John Parkinson, is of active-region material and shows the lines of a number of ionization stages. Particularly prominent are lines of Fe XVII, which indicate a gas temperature of around $2\text{--}3 \times 10^6$ K. A multigrid collimator was placed in front of the spectrometer so as to define an angular field of view of 3 arc minutes square on the Sun. Thus the radiation comes mainly from a single active region. Even lower temperatures would be found in parts of the corona remote from active regions. In fact active regions contain plasma which ranges in temperature up to about 7×10^6 K but falls short of the value reached during solar flares.

The other spectrum in Fig. 3.2(b) was obtained by the NRL group during a solar flare. Flares are sudden and dramatic releases of energy and we will discuss their properties in more detail in the next chapter. However, the contrast between active region and flare spectra is clear and is due to the large volume of high-temperature gas produced in the latter case. Whereas the highest ionization stage of iron observed in the active region spectrum was Fe XVII, stages up to Fe XXII are clearly present in the flare spectrum and we will see later that even the helium-like stage of iron (Fe XXV) is usually produced (see Section 3.1 for an explanation of this notation). This indicates that the gas is being heated to temperatures of around $25\text{--}30 \times 10^6$ K.

We will see in later chapters that plasmas with temperatures as high as $\sim 2\text{--}3 \times 10^8$ K exist in other parts of the Universe, namely in supernova remnants and in clusters of galaxies. However, it is in the corona, the atmosphere of our nearest star, that we are able to study such gas in the greatest detail. Although coronal lines are produced in the visible part of the spectrum, they are very difficult to detect because of the extreme brightness of the underlying photosphere. X-ray observations allow us to study the solar atmosphere without the need to use occulting discs or to wait for infrequently occurring eclipses. In the next sections we will discuss some of the results of these studies.

3.3 Magnetic fields and coronal structures

When the existence of high-temperature plasma was discovered in the outer parts of the solar atmosphere, it was assumed that the hot-gas distribution was spherically symmetrical, falling off in density radially. This simple model was supported by the fact that an assumption of hydrostatic equilibrium together with the measured gas-density gradient yielded a temperature similar to that obtained from visible observations of iron lines. If the corona were in hydrostatic equilibrium, then the pressure at every point would equal the weight of the overlying layers of gas at that point. Since, in a gas, pressure is related to temperature, it is possible to estimate the temperature from the known density gradient.

Even before the availability of solar X-ray images, the observations of the corona's visible light, carried out with the aid of occulting discs or during total solar eclipses, showed the complexity of the structures that existed in the coronal plasma (see Plate 1). Nevertheless, for many years it remained customary to think in terms of a general spherically symmetric coronal gas with active regions producing strictly local density enhancements here and there, like raisins in a pudding. The concept of a general corona in hydrostatic equilibrium was retained for as long as possible because of its extreme simplicity.

The true nature of the solar corona was revealed only when observations were made with imaging X-ray telescopes of the Wolter I type (see Chapter 2). Some of the best work in this area has been carried out by Vaiana and Krieger and their co-workers at AS & E. They undertook a series of rocket flights of their telescopes, which culminated in a set of three relatively long-duration observations carried out on the manned Skylab mission in 1973. This spacecraft was equipped with a highly accurate and stable solar pointing platform (the Apollo telescope mount), which allowed X-ray images of arc second quality to be obtained.

Two examples of these X-ray images are shown in Plate 7 together with an explanatory diagram. Since photographic film was used to register these images, there is very little information available about the spectra of the structures that can be seen. However, a combination of filters, the cut-off in the telescope reflectivity and the steepness of the spectrum with photon energy ensure that the observed energy range is 0.1–2.0 keV. The number and complexity of loop and arch-like structures is immediately obvious. It has been determined from a variety of other evidence that these structures coincide with lines of magnetic force, which break through from the solar surface. In this section we will discuss the larger—and somewhat cooler—features, leaving the active regions and bright points until later.

Coronal holes are among the largest and most dramatic structures visible in these images. They are regions of low-X-ray brightness, in which the magnetic field is directed radially outwards from the solar surface and the lines of force do not return to the photosphere. The way in which these regions are believed to form is illustrated in Fig. 3.3. Active regions are normally bipolar, that is they show positive and negative magnetic fields. The Sun rotates in the direction of the arrow with a period of approximately twenty-seven days. As it rotates, the magnetic polarity of the leading region is different in the northern hemisphere from that in the southern. When a pair of active regions are situated above and below the equator, as shown in Fig. 3.3, then a large region of uniform positive (or negative) polarity is formed, which crosses the equator. The field changes sign at the boundary of this region but, within the unipolar area, field lines are directed outwards and do not return to the solar surface. It was established by correlating the Skylab X-ray images with solar magnetic-field measurements that coronal holes are formed above these large unipolar regions. The material within a hole is cooler and less dense than in other parts of the corona. In addition, it has been known for some time that there is a continual outflow of material from the Sun, which is referred to as the solar

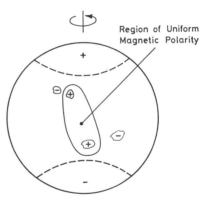

Region of Uniform
Magnetic Polarity

Fig. 3.3 Formation of a coronal hole. The appropriate alignment of two active regions above and below the equator permits the establishment of a large area of uniform magnetic polarity on the Sun

wind. The velocity and density of this outflowing material varies with time. Studies of a number of Skylab X-ray images revealed that the strength of flow of material in the wind was directly related to the number and area of coronal holes on the Sun. Thus material outflow takes place preferentially above these regions of open magnetic field. This discovery has been important to other branches of astronomy. We shall see later that material outflows, or winds, also exist in the case of several stars and that these winds have important implications for the X-ray emission from binary-star systems. Also, the arrival of solar-wind particles can strongly influence the outer layers of the Earth's atmosphere.

In addition to the open coronal holes, there are closed cooler regions called filament cavities. These are regions of low X-ray emission immediately surrounding quiescent filaments. Filaments are large arch-like structures which contain gas at a much lower temperature than that of the rest of the corona. Called *prominences* when seen at the edge or limb of the Sun, these regions of cool gas can exist for a month or more but on occasion only last for a few hours. Coronal regions immediately adjacent to these filaments, perhaps not unnaturally, show lower temperature and density than other parts of the corona.

Except in coronal holes, the X-ray images in Plate 7 both exhibit *limb brightening*. This is due to the existence of more hot material along an observer's line of sight when he looks towards the Sun's limb, compared with a line of sight which points at the disc. The brightening can occur because, unlike the photosphere, the coronal gas is transparent or thin to its own radiation. Hence the existence of more material along a line simply increases the quantity of X-radiation emerging from that direction. In fact, the reverse of the limb-darkening effect occurs in the photosphere, where radiation is absorbed after travelling a very short distance and thus only the radiation from the outer layers can be seen. Although the discovery of limb brightening in early solar X-ray images seemed, to some degree, to support the existence of a spherically symmetrical corona, detailed analysis of images like those in Plate 7 shows that this is not the case. Magnetic structures are revealed to control the location and disposition of the gas virtually everywhere in the corona. This is because the gas is an ionized plasma made up of negative electrons and positive ions. These charged particles can be contained by a magnetic field of adequate strength. Thus the charged particles are everywhere trapped in closed magnetic structures except in the case of coronal holes, where they flow out along open field lines.

The realization that coronal structure was completely dominated by the magnetic field pattern was a major discovery, made possible principally by X-ray observations. Although visible-light observations had for many years suggested the same finding, the need to observe only at the Sun's limb made progress difficult. X-ray observations made it possible for the first time to study the corona in front of the disc of the Sun.

3.4 Active regions, loops and bright points

We have already referred to the local density enhancements in the corona that occur in what are called *active regions*. Before discussing the X-ray output of these regions and its relevance to our understanding of the nature of the corona, it is worthwhile making some brief comments about solar activity.

The Sun has a small general magnetic field of about 1 Gauss in strength. It is also known to rotate differentially. Thus a feature at the solar equator will exhibit a twenty-seven-day rotation period, whereas at the poles the appropriate period is thirty-two days. Since the magnetic field exists in the surface layers, the differential rotation will wind up and concentrate the magnetic field lines in the manner shown in Fig. 3.4(a). The frozen-in field lines are moved closer together by the shearing effect of differential rotation and the magnetic field strength is thereby increased. While this shearing of the magnetic field is going on, convective motion of gas is continually occurring below the surface of the photosphere. This motion can twist the magnetic field lines into rope-like structures, which further increases the field strength. The convective motions can also cause the twisted field lines to break through the surface of the photosphere as illustrated in Fig. 3.4(d). It is this looped-field configuration breaking through the surface, which gives rise to active regions in the solar atmosphere. The reversal of the field direction in the northern and southern hemispheres leads to the preceding and following active-region fields being reversed as shown—recall the importance

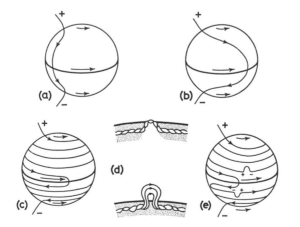

Fig. 3.4 Differential rotation and formation of active regions. The winding up of the solar magnetic field due to differential rotation are shown in (a), (b) and (c) and the resulting formation of bipolar active regions in (d) and (e)

of this in the formation of coronal holes as discussed in the previous section.

The existence of localized bipolar magnetic field regions, of the kind described in the previous paragraph, gives rise to the many observed manifestations of solar activity. However, a complete discussion of these is beyond the scope of this book. Perhaps the best known active-region phenomenon is that of *sunspots*, which are the darker regions that appear on the Sun's surface. Their number and distribution allow us to characterize the rise and fall of solar activity in its eleven-year cycle. However, we are more concerned here with the effect of the magnetic loops on the coronal gas and its X-ray emission. It is important to remember that while the energy of the Sun in general comes from the nuclear fusion of hydrogen in the core, it is the combination of the differential rotation and the solar magnetic field that supplies the energy for solar surface activity and the channels or structures through which this energy passes into the corona.

Returning to Plate 7, a detailed analysis of the images shows that the coronal gas in the neighbourhood of active regions is organized into very many loops or arches. This is not easy to see directly in the two images shown in Plate 7. However, a large number of images taken at a variety of different exposures can show the required detail. Loop sizes extend from arc minutes in length and a few arc seconds in cross-section down to sizes that cannot be resolved with the X-ray telescopes (angular resolutions of 1 arc second) that have been available up to now. The thin solid lines in Plate 7 indicate the location and approximate scale of the coronal loops. These loops generally run from zones of opposite magnetic polarity within a single active region. However, the Skylab observations showed for the first time the existence of loops that connected different active regions to each other, a feature of solar activity that is still not properly understood.

We can conclude from studies of many solar X-ray images similar to those in Plate 7 that most of the hot coronal plasma which appears in the plates is contained in loops or closed magnetic structures.

From a knowledge of the sizes of loops and from measurements of the X-ray output, it is possible to make estimates of the plasma density, which is found to lie in the range 10^8–10^{10} particles per cm³. The temperatures deduced from Skylab images taken with different filters are in the range 2–3×10^6 K. Unfortunately, the spectral resolution of the filter–film combination is inadequate for the resolution of individual spectrum lines and so it is not possible to obtain really good plasma temperature measurements. Bragg crystal spectrometers show the presence of emission lines from ion species that indicate active-region temperatures of up to 7×10^6 K, but these instruments do not yet have adequate spatial resolution to show where this hotter material is located within the active-region structures. We will return to a discussion of the temperature distribution in loops later.

The lengths of loops lie in the range 10^3–10^5 km and the amount of energy contained in the hot gas in loops seems to depend inversely on loop size and age. Thus the larger and older loops contain less energy than the younger and smaller loops. This information has been obtained by studying the same active region on successive solar rotations. The lifetimes are, in fact, much longer than would be possible if an initial energy input was merely being radiated away in the X-ray band or fed to the footprints of the loops by thermal conduction. Hence the presence of a continuous source of energy is indicated.

It is reasonable that the coronal material should, in general, be confined to a number of separate and distinct active-region loops. This is because it is very difficult for plasma to flow across magnetic field lines, but very easy for it to flow along them. Thus, the thermal conductivity or the ability of the gas to conduct energy is very high along the length of a loop but very low along a loop radius.

Although studies of single loops are difficult because of the small sizes involved, some attempts have been made to deduce the properties of individual structures from the observations. Models of loops have been made which have properties that are in good agreement with those observed. In particular, satisfactory relations between temperature, pressure, length and radius have been constructed. The required continuous supply of energy must be deposited along the entire length of the loop.

This constraint on the energy input to loops is of the utmost importance since it bears directly on the question of how the coronal gas can be supplied with sufficient energy to raise its temperature to above 10^6 K. At one time it was believed that the energy came from the photosphere in the form of sound waves generated by the convective turbulence just below the Sun's surface. However, sound energy could not be deposited uniformly over the very large distances in coronal loops. Instead it seems more likely that the magnetic field structures act as guides for magnetic waves and that the coronal energy is supplied in this form. A great deal more work is required in this area but it appears likely that X-ray observations of active-region loops may at last permit us to solve the long-standing problem of how the corona is supplied with energy and thus heated to 1,000,000-degree temperatures.

The remaining features of interest in Plate 7 are called *coronal bright points*. These were first discovered in the Skylab X-ray images. They appear as spots of around 20 arc seconds (14,000 km) in diameter and often have 5 arc second bright cores. Although found at all solar latitudes, they emerge preferentially in mid-latitudes where active regions are created. In fact two kinds of bright points exist. Those found in active-region latitudes (that is between 35° N and 35° S) have typical lifetimes of thirty hours while the more generally distributed features live for only eight hours. A comparison of the X-ray bright points, with optical magnetic field measurements in the same regions, shows the bright points to be small bi-polar magnetic regions. They, therefore, appear to be smaller editions of the familiar active regions which we have already discussed. Thus the X-ray observations have led to the discovery of a new form of solar activity.

4 · Solar-flare X-rays

4.1 The nature of solar flares

Solar flares are violent outbursts of electromagnetic radiation. They appear suddenly, are short-lived and occur over a wide wavelength range, namely from 0.001 Å (10 MeV, 10^{21} Hz), which is in the gamma-ray range, through the entire spectrum to 10^4 m (3×10^5 Hz) in the radio range. The individual photon or quantum energy is not significant at radio wavelengths but is, of course, all-important for gamma rays. Some features of a solar flare can develop and decay in time scales of seconds, while others remain visible for many hours.

Flares are of interest for a number of reasons. First, their origin is not yet completely understood although it is almost certain that the flare energy is supplied in some way by the magnetic field. A full understanding of the energy storage-and-release mechanism may have implications for power generation by nuclear fusion. Second, the electromagnetic and high-energy particle emissions associated with flares have a significant effect on the atmosphere of the Earth that is of importance for communications and possibly for the weather patterns. Third, the availability of a vast output of energy over a wide range of wavelengths has stimulated important work in atomic and radiation physics. Finally, it is known that flare-like outbursts are observed in certain stars and there are likely to be similarities between the two phenomena.

Some impression of the wide range of electromagnetic phenomena observed in a solar flare may be obtained from the chart in Fig. 4.1(a). Wavelength (λ) units are shown throughout in centimetres on a logarithmic scale. Thus the exponent or power of 10 is plotted rather than the actual numerical value of the wavelength. Underneath the centimetre scale, the units of wavelength more appropriate to a particular spectral range are also shown. Photon-energy and frequency units are also indicated, as are the usual names given to the

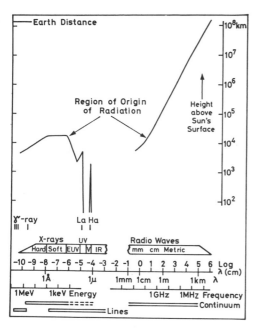

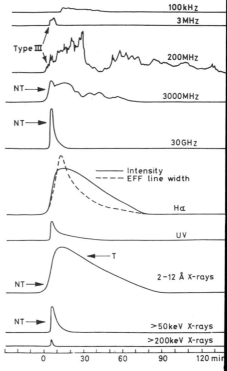

Fig. 4.1(a) *(above)* A chart showing the range of radiation wavelengths produced during solar flares and the approximate location of the region *(right)* of origin in the solar atmosphere (on a logarithmic scale)

Fig. 4.1(b) Plots of solar-flare intensity against time for a variety of wavelengths

different spectral ranges. Radiation may be emitted mainly as continuum, mainly in lines or as a mixture of both and this is also shown. Finally, the region of the solar atmosphere from which a particular radiation generally originates is also shown. It is interesting to note that some of the radio emission can be produced quite close to the Earth itself.

In studying solar flares, it is quite often useful to plot the intensity of the radiation in a particular spectral range against time. Such plots are called *light curves*. A number of these are shown for the different spectral ranges in Fig. 4.1(b). The letters NT indicate that a non-thermal process is involved in the production of the radiation. Such processes usually include beams of electrons where the velocity distribution is quite different from that of particles in a hot gas, that is, it cannot be simply characterized by a single temperature.

We will return to a discussion of these processes in a later section. It is significant that the non-thermal phenomena are more impulsive —they rise and fall more quickly—whereas the thermal emissions, which originate in hot-plasma processes similar to those discussed in the previous chapter, are more gradual in their behaviour. In addition to the radiation output of a flare, a substantial part of the flare energy appears in the form of accelerated electrons and photons. Many of these can escape from the Sun and are detected near the Earth by satellites equipped with particle detectors.

Before the advent of space astronomy, flares were studied from the Earth mainly at visible and radio wavelengths. Much of the visible data has been obtained in the light of a particular line, the strong Hα line emitted by hydrogen at a wavelength of 6563 Å. An example of a flare in Hα light is shown in Plate 8. These photographs, taken at the Lockheed Solar Observatory which was then located near Los Angeles, show a very rapidly developing flare which starts out in the neighbourhood of an active-region sunspot. Notice the rapid growth of the emitting region. This is in fact an explosive flare which is marked by a particularly rapid growth in flare area (Plate 8 shows an explosive flare). The period of most rapid growth in Hα intensity is called the flash phase. Flares which occur at the solar limb can be seen in projection and so their height above the solar surface can be measured. Typical values reached are less than 10^4 km. Flares are classified by their area at the time of maximum brightness. The classification goes from s (small) through 1, 2 and 3 to 4 for the larger flares. Hα-flare durations vary from minutes to hours. Flare-occurrence rates follow the number of sunspots, and therefore the number of active regions on the disc at any instant, but there are bursts of flare activity for short periods within solar cycles.

Solar flares occur only within active regions or in bright points which are, in fact, simple active regions. The frequency of occurrence of flares and their flare size or importance are related to the pre-flare rate-of-change of the magnetic field and to the magnetic complexity of the active region. The most important observation so far made, however, is the discovery that flares generally begin as two regions which brighten simultaneously on opposite sides of the region's magnetic neutral line. This is illustrated in Fig. 4.2 where the neutral line, the boundary between regions of opposite magnetic polarity, is shown superimposed on contours of the magnetic field of the active region. The hatched regions indicate the locations of the flare.

We will see later that the X-ray emission occurs high in loops which cross over the neutral line.

Even though the flaring regions are bright and a great deal of energy is released, it is usually impossible to see the visible flare against the brightness of the solar disc, unless a narrow spectral range including the strong Hα line is employed. However, on rare occasions, flares can be seen against the disc in ordinary light. These so-called white-light flares, first discovered by Carrington in 1859, are unusual events which may include non-thermal emission, associated with high-energy electrons or protons.

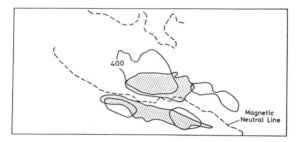

Fig. 4.2 The location of a solar flare illustrated by the hatched regions on opposite sides of the magnetic neutral line. The contour lines denote magnetic field intensity (Gauss).

A wide variety of radio-wave phenomena can be observed with solar flares. Examples of radio bursts on a number of different frequencies are illustrated in the upper part of Fig. 4.1(b). Some of these are gradual, and their light curves rise and fall slowly. Such events are probably due to the heating and cooling of coronal gas.

At a frequency of 200 MHz, examples of type-II and type-IV bursts are illustrated. The type-III bursts are believed to be due to the passage of accelerated electrons out through the solar corona. Travelling at around 10 per cent of the speed of light, these electrons force the local ions and electrons to oscillate everywhere along their path. The local electrons in turn radiate at a frequency characteristic of the local particle density, which is called the plasma frequency. Bursts of radio emission are seen at successively lower frequencies as the original high-energy electrons move out through the solar atmosphere. Because of the change in burst frequency, the type-III bursts are called drifting bursts.

The impulsive microwave radio bursts can rise and fall in times of the order of seconds. They are clearly too rapid to be due to heating and cooling of coronal gas and another explanation must be sought. From the previous chapter it is clear that magnetic-field loops exist in the solar atmosphere above active regions. The existence of type-III bursts points to the presence of fast electrons. If we investigate the motion of fast electrons in closed magnetic structures, we discover that they execute a helical motion around the magnetic lines of force. In so doing they undergo centripetal acceleration and so give off electromagnetic radiation known as synchrotron radiation (see Fig. 4.3). The electrons travel rapidly to the bases or feet of the closed magnetic structures, where they are absorbed by the dense matter and emit X-rays (see Section 4.2). However, on their way down the magnetic loops, they emit the microwave radio bursts.

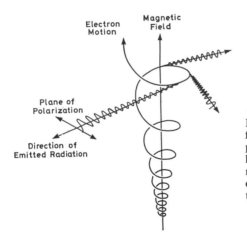

Fig. 4.3 Radio emission from the synchrotron process. In executing helical motion around magnetic field lines electrons emit synchrotron radiation

In this section we have discussed the nature of the flare as observed at visible and radio wavelengths. The Hα pictures, although very detailed, are unable to give us adequate insight into the physical process taking place. However, they do demonstrate the importance of the active region's magnetic field and, in particular, the relevance of the magnetic neutral line to the manner in which the flare occurs. The radio observations demonstrate conclusively that the flare process leads to the acceleration of large numbers of high-energy electrons. These electrons play a significant role in the flare X-ray production.

4.2 X-rays from heated flare plasma

The Hα emission from flares, which was discussed in the previous section, occurs mainly in the photosphere. The temperature of the flaring photospheric material does not rise above 10,000 K. However, the gas temperature is increased dramatically in the corona; it can reach values as high as 30–40×10^6 K in small regions at the flare site. At these temperatures, a considerable flux of X-rays is emitted and it is this X-radiation that we now wish to consider.

Two different X-ray flares, which develop at the solar limb, are shown in Plate 9(a, b). The position of the limb is shown by a solid white line. Both images were obtained by the AS & E X-ray telescope on Skylab. The flares occur in loops and usually begin near the top of a loop. The loop 'footprints', which are in the photosphere, appear bright in Hα and are on opposite sides of the magnetic neutral line. No Hα emission can be seen at the top of the loop, since a gas that is hot enough to produce X-rays cannot emit Hα; no neutral hydrogen atoms can exist at temperatures suitable for X-ray emission because bound electrons are removed from these atoms by collisions with the energetic free electrons in the gas. A variety of different flaring loops have been seen and we will return to a discussion of these later. The X-ray energy range of these and other Skylab images is about 0.2–2 keV, though the instrument response within this band is somewhat affected by the type and thickness of filter in use at the time of the observation.

A striking feature of flare observations at around 1 keV is the dramatic increase in plasma temperature that is observed. This was discussed briefly in relation to the spectrum obtained with a Bragg spectrometer by the NRL group (Chapter 3, Fig. 3.2). Spectra of this kind show the gas temperature rising rapidly to temperatures above 20×10^6 K in a few minutes or less. Thus, the occurrence of a flare must lead to a very rapid release of energy into the coronal plasma—probably at the top of a magnetic loop associated with an active region. It is important to remember, however, that the rise and fall in X-ray intensity at around 1 keV is gradual when compared with the rate of change of intensity at 20 keV and above (see Fig. 4.1(b)).

Very many emission lines are produced by the hot plasma, as was shown in Fig. 4.3. The very existence of highly stripped ions (e.g. Fe XXV—iron stripped of 24 electrons) in the gas gives a rough indication of the very high temperatures reached. However, the

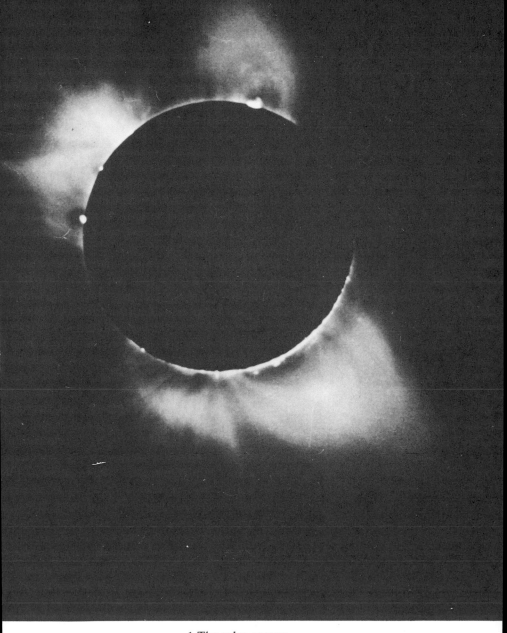

1 The solar corona

2 The Copernicus satellite

3 *(left)* The 'WAC Corporal' of the Jet Propulsion Laboratory, the world's
first successful sounding rocket, and Frank J. Malina who directed its
design, construction and testing. (White Sands Proving Ground,
New Mexico, U.S.A., September 1945)

4 *(right)* The British Skylark vertical sounding rocket leaving the
launcher at Woomera

5 *(above)* The Ariel V X-ray satellite
6 *(below)* The Einstein X-ray observatory

FILTER 3
1 sec.

FILTER 3
1 sec.

24,1307:44 UT   25.0207:46 UT   (a)

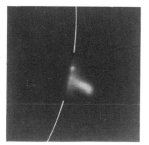

FILTER 3
0.5 sec.

FILTER 3
0.5 sec.

0805:09 UT   1257:57 UT   (b)

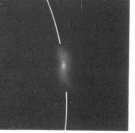

0727:56 UT   0903:05 UT   1039:26 UT   (c)

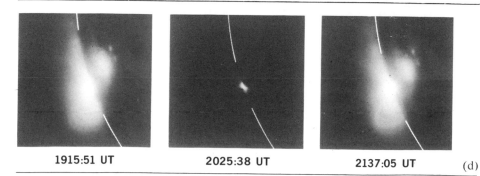

1915:51 UT   2025:38 UT   2137:05 UT   (d)

N

W

2 arc min.

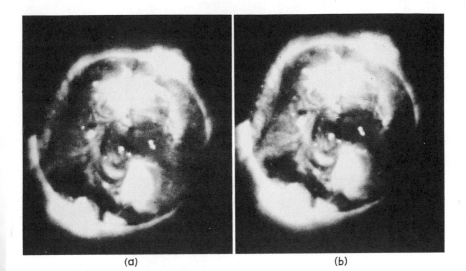

(a)   (b)

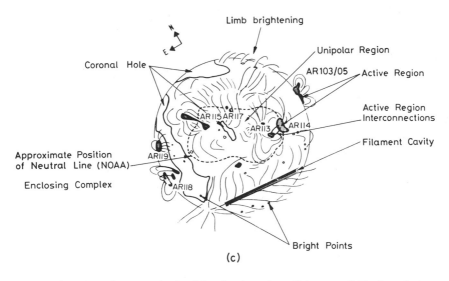

(c)

7 Solar X-ray images obtained by the American Science and Engineering
X-ray telescope on the Apollo telescope mount: (b) is a longer exposure
than (a), and (c) provides a key to the types of features observed

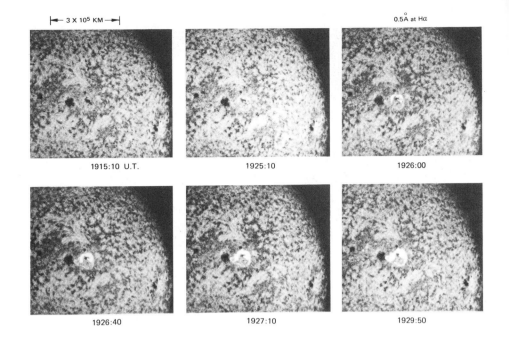

3 X 10⁵ KM

0.5Å at Hα

1915:10 U.T.   1925:10   1926:00

1926:40   1927:10   1929:50

8 A solar flare in hydrogen-α light photographed at the
Lockheed Solar Observatory. 11 August 1960

facing page

9 Four X-ray flares seen at the solar limb by the Apollo telescope
of American Science and Engineering: (a) 24–5 September 1973,
(b) 15 January 1974, (c) 4 December 1973, (d) 11 January 1974

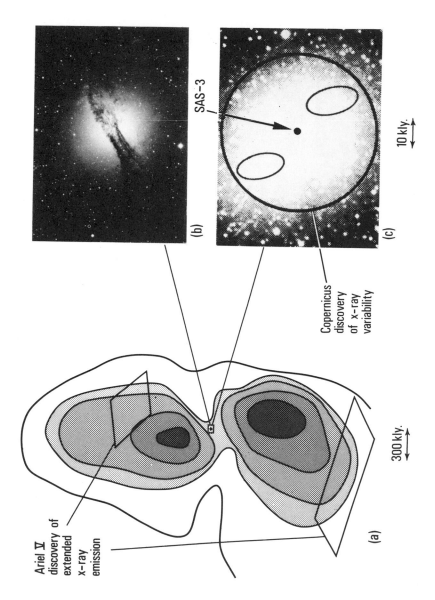

Ariel V
discovery of
extended
x-ray
emission

300 kly.

(a)

(b)

SAS-3

Copernicus
discovery
of x-ray
variability

10 kly.

(c)

10 Three views of the giant radio galaxy Centaurus A

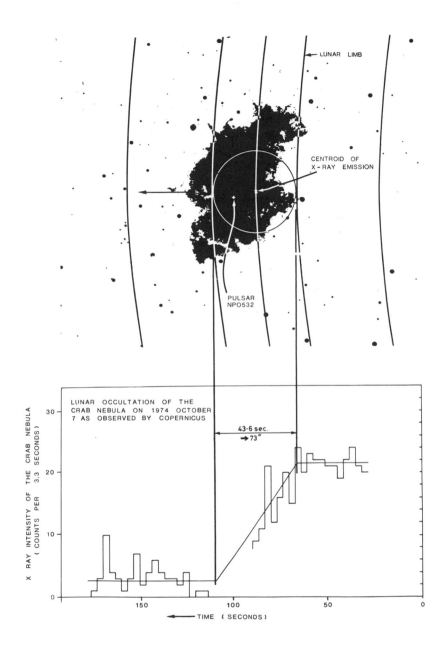

11 The Crab Nebula occultation experiment carried out with the Copernicus satellite

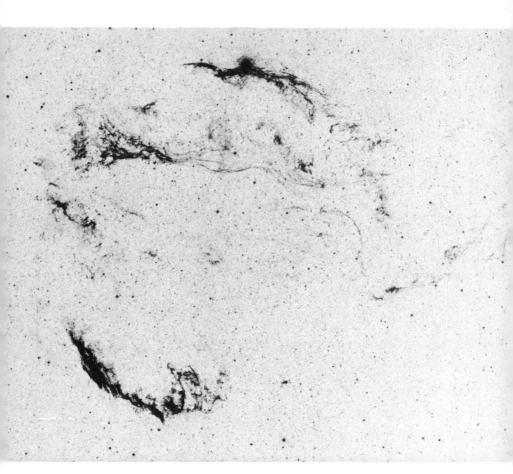

12 The Cygnus Loop or Veil Nebula

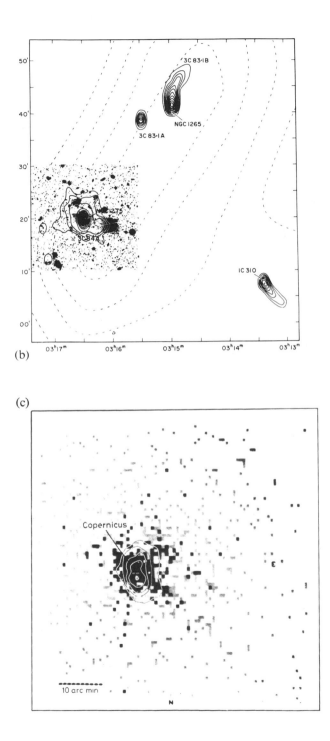

(b)

(c)

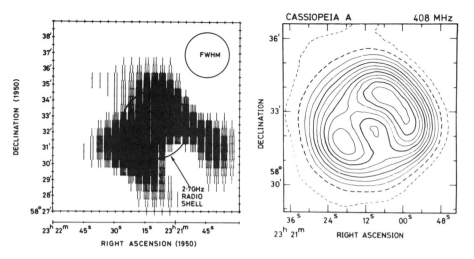

13 An X-ray map of the Cassiopeia-A supernova remnant

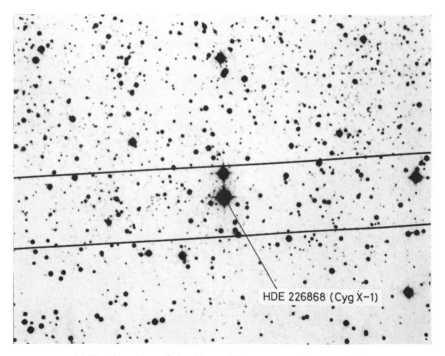

14 The location of the binary X-ray source Cygnus X-1

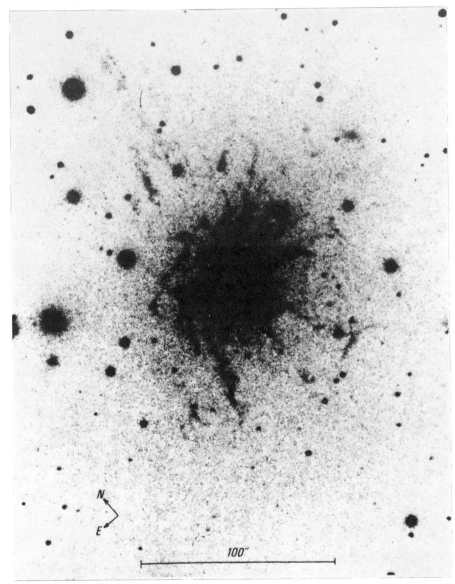

(a)

15 Radio, optical and X-ray views of the Perseus Cluster: (a) is the
galaxy NGC 1275 at the centre of the cluster, (b) is a map of a radio
emission from the cluster with NGC 1275 inset, and (c) shows the
X-ray emission from NGC 1275 from the cluster

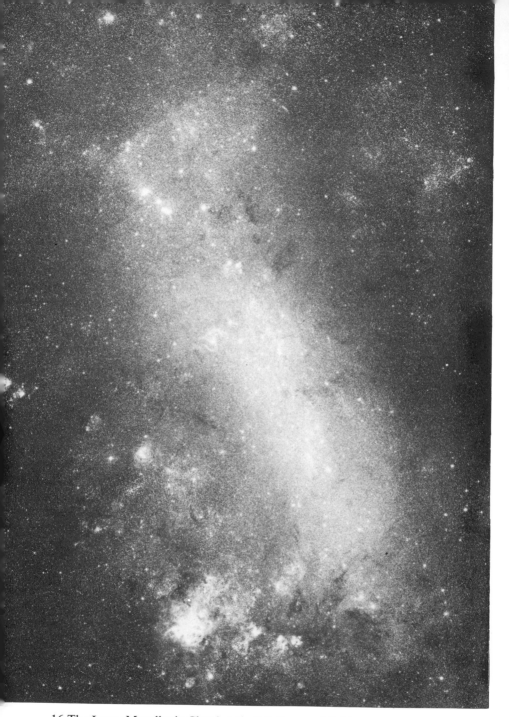

16 The Large Magellanic Cloud, only 170 thousand light-years away,
is the largest of a number of small galaxies which are satellites of our
Milky Way. It is of a type called Irregular, so named because it does
not exhibit the prominent spiral structure and symmetric central bulge
common to galaxies like ours. More than 100 X-ray sources have been
discovered, with the Einstein observatory, in the Magellanic Clouds

17 The Small Magellanic Cloud, also one of the satellite galaxies of our
Milky Way, is about 200,000 light-years away. It is easily visible to the
naked eye from the southern hemisphere. It lies quite close to its
bigger (and slightly closer) companion, the Large Magellanic Cloud

18 The spiral galaxy M 51 and irregular galaxy NGC 5195.
Photograph from the Hale Observatories

measurement of specific line intensities can give a more accurate estimate of temperature—recall that emission lines are produced in collisions in which the energetic free electrons in the plasma excite bound electrons in ions to higher energy states. The bound electrons emit X-ray photons in falling back to ground levels. The ratio of the intensity of two different spectrum lines with different transition energies depends sensitively on the plasma temperature. Other emission lines, chiefly those emitted by helium-like and beryllium-like ions (i.e. ions with 2 or 4 electrons remaining), allow estimates of particle density to be made. Although this kind of measurement is only now being developed for solar-flare observations, there are indications that remarkably high densities may exist, particularly at the very centre of the flare region. Yet further lines, the so-called satellite lines, produced in the spectra of helium-like ions, can indicate whether or not the plasma is excessively ionized for the measured temperature. This kind of measurement has important implications for the understanding of how energy is being released into the gas. Thus the use of high-resolution X-ray spectrometers can tell us a great deal about the properties of the hot gas. Unfortunately, the spatial resolution that can be achieved with such spectrometers is not yet very high. In particular it cannot match the arc second performance of the imaging X-ray telescopes. These, however, cannot resolve the individual spectrum lines. It is to be hoped that the spectrometers flown on the Solar Maximum Mission satellite early in the 1980s will go some way towards allowing us to deploy the very powerful spectroscopic techniques described above to observe the hot plasma produced in solar flares. At the time of writing, however, most of the progress being made in understanding flare plasma is coming from studies of the Skylab flare images.

A further variety of X-ray-flare images is shown in Plate 9(c, d). The Skylab studies have revealed two basic kinds of flare loop. Firstly, there are compact low-lying loops the maximum height of which are less than 25,000 km above the solar surface. Loop lengths are not more than 40,000 km. On occasion, because of the way in which a loop is located relative to the disc, the loop structure cannot easily be discerned and the flaring regions appear as small spots of approximately 7,000 km in diameter. The volumes of hot plasma involved in these events are in the range 10^{26}–10^{27} cm³ and the average particle densities lie in the range 7×10^9 to 2×10^{11} particles per cm³. These values are about ten times greater than those generally found in the corona and the additional material must be supplied to the flaring loop from elsewhere. In the case of the compact low-lying loops, it is believed that material streams up the loop from the photosphere. This upward-streaming activity has been called evaporation, since it is probable that the energy released initially in the flare is transported down the loop to the photosphere. Here, it 'evaporates' material, which is heated and streams up the loop emitting X-rays.

The second type of hot-plasma event occurs in large loops with maximum heights in the range 40,000–80,000 km above the surface. Hot-plasma volumes for these events are around 10^{29} cm^3 and densities are still ten times greater than the coronal values. It is unlikely that 'evaporation' from the photosphere could supply structures as large as these. It is believed, instead, that energy is released into existing filaments or prominences. The material in these structures is heated, giving rise to the X-ray emission. This suggestion is supported by the frequently observed phenomenon of disappearing filaments. Recall that filaments (called prominences on the limb) are long threads of cool gas supported high in the corona by magnetic field structures. The disappearance of the filament simply means that the structure stops emitting Hα because it has been heated to X-ray-emitting temperatures by the flare-energy release.

Thus, in both kinds of event, a pre-existing active-region loop is filled with hot material at a temperature and density substantially greater than that of a non-flare loop. Unfortunately it will require better time and wavelength resolution than have been available in instruments used so far to understand the way in which energy is released into the loop. It is already apparent that the energy source must be in the magnetic field but the details of how the energy is stored and released are still obscure. We will summarize the models for this mechanism that have already been proposed in Section 4.4, but first it is necessary to describe the X-ray emission from the high-energy non-thermal electrons which are accelerated during solar flares.

4.3   X-rays generated by accelerated electrons

In addition to the hot-plasma X-rays, which are emitted during flares in the energy range 0.1–10 keV, there is an impulsive component of higher-energy X-rays. X-rays with energy above 20 keV were detected in a balloon flight carried out by Winckler and Peterson in the USA in 1959. Studies in this energy range, using scintillation detectors, have continued up to the present time, but increasing use is being made of satellites for the long exposures needed to detect a large sample of events.

Bursts at higher X-ray energies are generally more impulsive than the lower-energy bursts emitted by hot plasma. In addition, the high-energy bursts occur simultaneously with—and have similar time profiles to—the impulsive microwave radio events discussed in Section 4.1, and we should recall that these microwave bursts were thought to be due to synchrotron radiation emitted by beams of high-energy electrons spiralling around the magnetic field lines in coronal loops. Thus it is immediately worthwhile to search for explanations of both phenomena that include the same high-energy electrons. The known electron energies and magnetic field strengths make it impossible for the X-rays to be produced by the synchrotron process.

Observations of the X-ray-burst sizes and locations on the Sun have been carried out by Oda and his colleagues in Japan using a modulation collimator. These measurements showed that the burst spatial extent was less than 30 arc seconds and that the X-ray-burst site was coincident with the site of an accompanying microwave burst. The optical cores and (when observed) the white-light spots that are arc seconds in size, have a brightness development which is usually well correlated in time with the X-ray burst. The hard X-ray bursts themselves occur during the Hα-flash phase of the flare but reach peak intensity about 1 minute before the Hα peak. Typical bursts rise to a peak in one to two seconds while the fall times are around ten seconds. However, recent observations have shown that many hard X-ray bursts break down into individual spikes of about one-second duration. While the continuous emission from a hot plasma exhibits an intensity that depends exponentially on photon energy ($I \propto \exp{(-E/kT)}$), the spectrum of the hard X-ray bursts is better represented by a power law in energy, which has the form $I \propto E^{-\alpha}$. Values of the exponent α are usually found to be around 4.

All of the above properties suggest that a different emission mechanism may be involved from that which we have encountered in the case of a hot plasma. Studies of flares in the photon-energy range 1–10 keV have led us to believe that emission in this interval is thermal and is due to the interactions of a randomly moving assembly of electrons and ions whose energy distribution can be characterized by a temperature. In the case of the hard X-ray bursts, there are a number of indications that a non-thermal process is the cause. First, the measured hard X-ray spectra observed in solar flares extend up to photon energies above 500 keV. While it would be possible to explain the power-law spectral shape by a gas with elements at many different temperatures, a maximum temperature of around 2×10^9 K would be required. This seems impossibly high. We know from the X-ray images that the hot plasma is contained in magnetic loops high in the corona. Magnetic fields of more than 1,000 Gauss would be required to exist in these loops to contain such hot gas and there is no evidence to suggest that magnetic fields of this strength exist in such large volumes of the corona. Second, hard X-ray-burst rise-and-fall times of less than one second have been observed. It would be very difficult to achieve such rapid changes in the radiation output of a hot plasma. Third, hard X-ray bursts coincide in time with microwave bursts and with the short-lived white-light outbursts and we believe that both of these phenomena are due to the interactions of non-thermal high-energy electrons and protons. Finally, substantial fluxes of particles are observed to leave the Sun and flow towards the Earth, where they are detected by Earth-orbiting and interplanetary satellites. Thus there is good evidence that the hard X-ray bursts are caused by accelerated non-thermal particles. We will now consider how the high-energy photons are produced.

Recall that the continuous spectrum of X-rays from the heated coronal plasmas is mainly produced by Bremsstrahlung—radiation caused by the deceleration of electrons as they interact in the electric fields of ions. We have seen the evidence for the existence of high-energy (up to several million electron volts) electrons during flares. These electrons are accelerated during the early part of the flare and are then guided down the magnetic field lines of the loop in which the flare is taking place. When the electrons reach the footpoints of a loop, they find themselves just above the photosphere in a region of very high particle density. The electrons are brought to rest in these regions by collisions with ions but in being slowed down they emit large bursts of X-rays by the Bremsstrahlung process. As remarked above, the spectra of these bursts are quite different from that of a hot plasma; they have power-law spectra which are most likely to arise from accelerated electrons whose energy distribution is also in the form of a power law.

There are some difficulties with this simple picture that remain to be cleared up. For example, it is not clear whether the X-rays are radiated as the electrons move through a region above the photosphere (the thin-target hypothesis) or whether they emit X-radiation only as they are finally brought to rest in the photosphere (the thick-target hypothesis). These two X-ray-production modes have different implications for the electron-acceleration process. In the thin-target case, it is implied that the electron-acceleration process stops at the peak of the hard X-ray emission. If the target is thick, then electron acceleration must continue throughout the duration of the X-ray event. Recent observations from satellites by Kenneth Frost in the United States and by Van Beek in the Netherlands suggest that each hard X-ray burst is composed of a series of repeated impulses lasting for around one second or so. This would imply repeated electron-accelerating impulses and would thus favour the thick-target model for hard X-ray bursts.

There is another, more serious, difficulty which has plagued the subject of solar-flare hard X-ray bursts almost from its inception. We have pointed out the remarkable simultaneity and similarity of time profile that exists between the hard X-ray and the microwave radio bursts. It is highly probable that these microwave bursts result from the synchrotron radiation emitted as electrons spiral in the magnetic fields of coronal structures. It is attractive to suppose that these same electrons proceed down the loops to the photosphere where they produce the X-ray bursts. However, even with the most favourable assumptions for the values of coronal magnetic field, electron-energy spectrum and Bremsstrahlung target density, the production of the observed X-rays by the Bremsstrahlung process requires 10^4 times more accelerated electrons than does the production of the observed microwave-radio burst by the synchrotron process. The explanation generally accepted for this discrepancy is that there are many mechanisms operating to absorb the microwave radiation in the corona and while the full microwave-burst flux may originally have been produced, only a small fraction of it escapes from the corona and is observed at Earth. This has never been completely convincing and there remains a problem to be resolved in this area.

4.4 Towards understanding solar flares

Before discussing the possible explanations for the occurrence of solar flares it is a good idea to summarize the observed features, particularly X-ray features, which a flare model must explain. The flare phases to be explained may be discussed under three headings.

Prior to the flare there appears to be some rearrangement of the magnetic field in the active region. This can be seen through some pre-flare thermal X-ray emission, which indicates a small degree of plasma heating. There is also some non-thermal ultra-violet emission and some evidence of filament activation.

The main flare exhibits a quasi-thermal phase in which there is a rapid increase of the coronal plasma temperature in the flaring loop. This temperature rise leads to a dramatic increase in 1–10 keV X-ray flux. Energy is conducted down the loop and heats the chromosphere and photosphere at the loop footpoints to around 10^4 K.

In addition the flare has an impulsive phase, during which electrons and protons are accelerated to many hundreds of keV. This acceleration coincides with the flare-flash phase and it may therefore occur right at the beginning of the flare or it may be slightly delayed. The accelerated particles lead to the production of non-thermal X-ray bursts, microwave radio bursts and the presence in interplanetary space, after the flare, of accelerated particles. These particles can, on their way out of the solar atmosphere, give rise to type-III radio bursts which are the result of induced oscillations of the coronal plasma.

Problems which any flare model must address are:

1. The source of the flare energy; a total of around 10^{26} J is contained in a large flare. J is the symbol for Joules.
2. The storage of this energy in the solar atmosphere.
3. The rapid release of the stored energy.
4. The gas-heating and particle-acceleration mechanisms.

These are the central questions concerning the nature of solar flares, which X-ray observations can help to resolve. However, convincing answers are not yet available, although almost all flare models suggest a substantial role for the magnetic fields in the active region.

Two general categories of model have been proposed. The first type envisages a passive role for the magnetic field. Here the magnetic field stores and guides energy, which is present in some other form. In the second type of model the magnetic field is itself a primary source of energy, in addition to acting as an energy guide.

Before examining these models it is worthwhile briefly discussing the plasma-wave phenomena known as Alfvén waves, following the work of the famous Swedish physicist Hannes Alfvén. An Alfvén wave is a coupled oscillation of both the magnetic field and the plasma. It is also sometimes referred to as a hydromagnetic wave. Alfvén waves propagate along the magnetic-field direction. It is primarily a disturbance in the magnetic field which moves at a velocity (v_A) that depends on the magnetic-field strength (B) and the particle density (ρ), or $v_A = B/ \sqrt{(4\pi\rho)}$. In the disturbance, since the ionized plasma is linked to the magnetic field, both field perturbation and plasma particles oscillate in the same direction. Since the ions are heavier than the electrons in the plasma, the latter move more rapidly than the heavier positive ions. This charge separation causes an electric field in a direction perpendicular to that in which the disturbance is being propagated and this field in turn causes a current to flow. Oscillations of charge in the photosphere could lead to the generation of Alfvén waves at the base of the corona. These waves will propagate in regions where the magnetic field is strongest and we have seen that strong magnetic fields exist in loop structures. Thus these loops can act as channels for the passage of energy into the corona.

Returning to the two categories of model, we will first discuss those models in which the magnetic field has a passive role. Piddington suggested that energy is stored below the photosphere in some unspecified manner. Mass oscillations are then said to start below active regions. These oscillations lead to the propagation of Alfvén waves along the field lines of a coronal loop and these waves carry enough energy to give rise to the observed flare. There is, in fact, no evidence to support the existence of regions of stored energy below the photosphere. Furthermore the supply of energy up through the footpoints of loops would conflict with the X-ray observations which appear to show that in many flares the primary site for energy release is at the top of a coronal loop.

Elliott has proposed a model in which energy is stored in the form of charged particles, mainly protons. These particles are accelerated and stored continuously in active regions and then released suddenly in one loop to produce a flare. This model is acceptable in being able to supply the energy; 10^{27} J could resonably be stored in this manner. However, there is a difficulty associated with the time taken to build up the required number of accelerated particles. At least 10^5 sec is required during which the stored protons would tend to lose energy by collisions with other particles and by interactions with inhomogeneities in the loop magnetic field.

Pneuman has proposed a model which is again based on Alfvén waves. It assumes that energy is being continuously dissipated in coronal loops, but is released more slowly than it is supplied. There is, therefore, a local build-up of energy in a loop, so that when the energy density in the gas exceeds that of the containing magnetic field, there is a sudden energy release and the occurrence of a flare. An energy of 10^{26}–10^{27} J could easily be collected in this way in a matter of hours for a modest flux of Alfvén-wave energy into a loop. This mechanism represents an effective way of producing the observed hot X-ray-emitting plasma but it cannot provide the acceleration to give high-energy particles. We are therefore compelled to consider models involving an active role for the magnetic field.

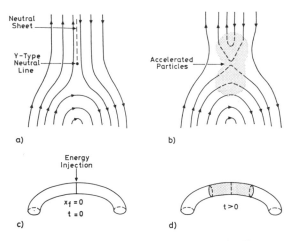

Fig. 4.4 A solar-flare model due to Sturrock. The rearrangement of lines of magnetic force (a) leads to the liberation of energy (b). The injection (c) and spreading (d) of energy in the loops is also illustrated

In models including use of the magnetic field as a source of energy, it is necessary to make 10^{26} J of energy available through the annihilation of part of the magnetic field of an active region. In order to release 10^{26} J in a volume of 10^{23} m^3, both reasonable figures for a large flare, around 25 Gauss in a field of original strength 500 Gauss must be annihilated. A number of requirements must be satisfied in order for field annihilation to make energy available. First, it must be possible for the magnetic field in which the energy is stored to be re-arranged or reconnected in a configuration with lower stored energy so that the difference may be made available. However, this configuration of the field must be stable against small disturbances, otherwise a build-up of energy would never be able to take place. Finally, the field rearrangement or reconnection must take place at a sufficient rate to release all of the energy produced in the change, in a time of between 100 and 1,000 sec. The important relationship between the magnetic neutral line and the occurrence of flares has led to the search for magnetic-field geometries that include a neutral line.

One suitable configuration, proposed by Sturrock, is shown in Fig. 4.4. It is a field arrangement such as might exist at the top of a coronal loop. In the rearrangement substantial electric fields are generated, which could give rise to accelerated particles that might both heat the plasma near the site of energy release at the top of the loop and also flow down towards the photosphere where the observed hard X-ray bursts are produced. The injection of energy into a loop is also shown schematically in Fig. 4.4. Other workers have discussed how energy is radiated and conducted away from the hot plasma.

This kind of model offers a useful explanation for the majority of flare phenomena, but a great deal of progress still has to be made before the solar-flare phenomenon is properly understood. The Solar Maximum Mission spacecraft, launched early in 1980, carries an array of ultraviolet, X-ray and gamma-ray instruments, which it is anticipated will finally lead to the solution of many of the problems associated with solar flares.

5 · X-rays from beyond the solar system

5.1 Satellite observations

We have seen in Chapter 1 how X-ray astronomy started with instruments carried on rockets; the first two sources discovered in our Galaxy, Sco X-1 and the Crab Nebula, were identified by observations of only a few minutes' duration, while the rocket was above the absorbing effects of the Earth's atmosphere. It was immediately recognized that longer exposures were needed and that these could be obtained only by the provision of instrumented orbiting satellites. Riccardo Giacconi, who had led the group which obtained the first evidence of X-rays from beyond the solar system, had approached NASA with a programme of exploration. In his words:

> Notwithstanding the paucity of information, I was young and enthusiastic enough to present to Dr Nancy Roman, Chief of the Astronomy Branch of NASA, a long-range program for X-ray observations extending over a decade . . . it was a delightful surprise to hear Dr Roman express the opinion that NASA might, in fact, be interested in considering an X-ray Explorer fully devoted to a search for celestial X-ray sources.

The satellite was eventually launched into an equatorial orbit in December 1970, with the aid of a Scout Rocket fired from a platform just off the coast of Kenya. After launch the satellite was named Uhuru, Swahili for 'freedom', in honour of the Kenyan Independence Day on which the satellite was launched. The satellite was aptly named, for the scientists involved were freed from the very limiting constraints entailed in the brief flight of a rocket. It became possible to observe X-ray sources for periods of many days, and with greatly increased sensitivity.

Uhuru was the first of a series of Small Astronomy Satellites (SAS) and its purpose was to perform a survey of the whole sky in the energy band 2–20 keV, in order to detect and determine the positions of sources that were as weak as 1/10,000 (10^{-4}) the strength of Sco X-1. Fig. 5.1 is an exploded view of the satellite showing the X-ray detectors and the star sensors, which are needed to determine the positions of any detected signals. Two detection systems were incorporated in the payload and these looked in opposite directions. The axes of the detectors were at right angles to the spin axis of the satellite, so that

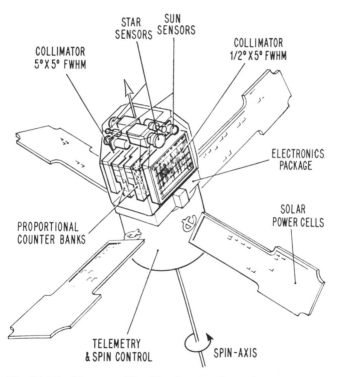

STAR SENSORS

SUN SENSORS

COLLIMATOR 5°X5° FWHM

COLLIMATOR 1/2°X5° FWHM

ELECTRONICS PACKAGE

SOLAR POWER CELLS

PROPORTIONAL COUNTER BANKS

TELEMETRY & SPIN CONTROL

SPIN-AXIS

Fig. 5.1 The Uhuru satellite. The drawing shows the X-ray detectors, star sensors and satellite systems for control. The detectors scan a region of the sky as the satellite rotates

during one rotation of the satellite the detectors scanned a band of the sky. Adjustment of the spin-axis direction allowed a scan of the whole sky. Conventional proportional counters were used to measure the X-ray signals and the fields of view of the detectors were restricted by collimators similar to those described in Chapter 2.

Giacconi's enthusiasm of seven years earlier was fully justified by the successes obtained with the Uhuru satellite. Within two years of the launch a catalogue was available to astronomers that contained more than 150 sources and their positions. This was an impressive achievement and allowed astronomers to begin the task of identifying the new X-ray sources with optical counterparts. We will see how important this is when, in later chapters, we deal with the detailed properties of individual X-ray sources.

By 1974, about thirty-five of the Uhuru sources had been identified with optical counterparts and the overall features of the X-ray sky were becoming apparent. We can see in Fig. 5.2 the distribution of sources from the Uhuru catalogue.

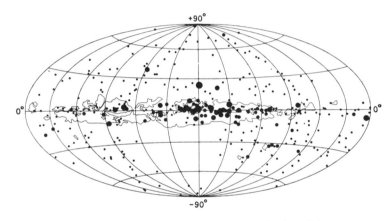

Fig. 5.2 A map of the X-ray sources. This plot of the X-ray sky in galactic co-ordinates shows that our galaxy contains many X-ray sources. The outline of the Milky Way runs from left to right at the centre of the diagram

This kind of map of the sky is called a plot in galactic co-ordinates; the centre of our Galaxy appears in the middle of the diagram. The contours correspond to the plane of the Galaxy, which is more familiarly known as the Milky Way. On a clear night we can see this as a broad band of diffuse light. With the aid of a small telescope this band can be seen to consist of large numbers of faint stars. It is clear from Fig. 5.2 that the X-ray stars are distributed mainly in the plane of the Galaxy but there is a concentration of bright sources in the direction of the galactic centre and a number of weak sources at high galactic latitudes.

Identification with optical counterparts is a crucial step in the understanding of X-ray sources, but the positions determined with the data from the Uhuru satellite were not precise enough to pinpoint a unique visible companion. For the strongest sources there could be as many as a hundred possible stars in the error box assigned to a particular X-ray source. For the weaker sources the signals were noisier and the error region could be as large as several square degrees. Optical astronomers were able to make some identifications on the basis of the unusual characteristics of a particular star—blue stars similar to the optical counterpart of Sco X-1 were obvious candidates. However, it was the discovery of regular variability in some of the Uhuru sources that provided the means of making identifications with certainty.

Two X-ray sources, one in the constellation of Centaurus and the other in Hercules, provided the most important discovery about the nature of some of the galactic sources made with the Uhuru satellite. The Centaurus source, named Cen X-3 (it was the third source to be discovered in the Centaurus region), was seen from one scan of the source—which lasted about 100 sec—to be pulsating in a regular fashion with a period of 4.8 secs. The other source, Her X-1, was found to be pulsating in a similar manner but with a period of 1.24 sec. When these sources were observed for several days, it became evident that in addition to the pulsations their X-ray emission was being cut-off completely on a regular basis. The pulsations were interpreted as the effects of the rotation of a neutron star and the regular eclipses indicated that the neutron star was itself in orbit about another star, the plane of the orbit being close to the line of sight.

The Hercules source is eclipsed with a period of 1.7 days and this feature provided the means for positive identification with an optical counterpart. A star with excess emission in the ultraviolet region was discovered in the vicinity of the X-ray source and shortly after, in 1972, this star was shown by William Liller—and independently by Neta and John Bachall—to have light variations with the same 1.7-day period. We shall see, in later chapters, how the measurement of regular variations in the signals from X-ray sources has become of great importance, for it provides ways of measuring the characteristics of collapsed stars and the conditions in their immediate environment. We will also see that the observations of the optical counterparts in these binary systems are essential to the understanding of the interaction between the two stars and the nature of the compact objects.

The other significant contribution of the Uhuru satellite was the discovery that some of the high-latitude sources were associated with clusters of galaxies. Although it was not possible to identify all the sources that had been discovered at high galactic latitudes, the stronger sources, which included the source previously identified with the galaxy M87, could be identified and studied. The M87 source was found to have an extended emission region, more in keeping with the cluster of galaxies to which it belonged. Three other clusters of galaxies also showed similar extended emission regions, which could be the result of emission from the individual galaxies of the cluster or a diffuse emitting region caused by material between the galaxies. We shall discuss these sources in more detail in Chapter 11.

The Uhuru satellite performed the important first survey of the sky in X-rays and made it possible for subsequent satellite experiments to undertake detailed observations of the sources that had been discovered. Fortunately, several such experiments had been planned and from 1971 to 1978 seven satellites (all with technical refinements to their instruments) continued the study of the X-ray sky. The remainder of this chapter is devoted to describing the two classes of X-ray source that have been extensively studied over the last eight years with these satellite instruments. The classes are naturally divided into those sources in our Galaxy and those which originate at the vast distances of the external galaxies and clusters of galaxies.

5.2 The galactic X-ray sources

Our Galaxy is a spiral structure of stars with a diameter of 25 kpc (kpc is the abbreviation for kilo parsec and is equivalent to 3.1×10^{21} cm or about 3,200 light years). There are about 10^{11} stars in the Galaxy and our solar system can be regarded as a tongue attached to the Orion arm of the Galaxy. On average, every 100 years or so one of the stars in the Galaxy explodes as a supernova. The remains of these explosions form the first class of source identified as emitting measurable amounts of X-rays. We have described the identification of the Crab Nebula as an X-ray source in Chapter 1. The most significant feature of these supernova remnants is that the X-ray emission does not vary. For this reason, the Crab Nebula is used extensively to calibrate instruments after they have been placed in orbit. It is indeed a fundamental characteristic of astronomical objects that the larger they are the less rapidly they can vary.

Although supernova remnants are an important class of galactic source, which we will discuss in detail in Chapter 6, the majority of X-ray sources in the Galaxy are known to arise from double stars. About half the stars in the Galaxy are to be found in the form of double stars; 30 per cent are seen as isolated stars and the remainder are multiple-star systems containing three or more stars. It is thus more common than not that a star is in a binary system, but it is *most unusual* for such a system to be a strong emitter of X-rays.

The discovery of regular variations in the X-ray signals from a number of sources has presented scientists with the opportunity to study many facets of these binary X-ray sources. By 1978 the periodicities observed in X-ray sources ranged from 0.033 seconds for the pulsar in the Crab Nebula to a possible 581-day period in the X-ray source associated with the star X Persei. This range is considerably wider than that seen in the radio pulsars, the reason being that only one, to date, is found in a binary system. It was immediately recognized that the discovery of regular, and therefore predictable, behaviour in X-ray sources would give a deep insight into the nature of these systems. More than a decade of intensive study on the X-ray emission of solar flares has so far failed to give a complete picture of the mechanisms responsible, mainly because of the unpredictable nature of the phenomena. In contrast, the astronomy of X-ray stars joins the established pursuit of recognizing regular patterns of behaviour which is basic to man's study of the Universe.

The discoveries of regular periods in such a large number of sources only became possible when X-ray instruments could be pointed for lengthy intervals at particular sources. The Uhuru satellite had of course made the first valuable discovery of the binary and pulsating characteristics of Cen X-3 and Her X-1, but the instruments of the Uhuru satellite could only register signals from a source as it flashed past in each twelve-minute revolution of the satellite. The next vital step was made, therefore, when the Copernicus satellite was launched in August 1972.

We have mentioned in Chapter 1 how the X-ray instruments on the Copernicus satellite were originally developed to attempt the detections of X-rays from the bright blue stars used in the observation of the interstellar material. In order to make the X-ray instruments as sensitive as possible, it was vital to reduce the background noise of the detectors, which was produced when high-energy cosmic rays interacted with the satellite to give showers of secondary particles. These background effects were minimized by using reflecting mirrors to collect the X-rays and then focussing them on to very small detectors. The small detectors presented a small cross-section for the cosmic rays and their secondary products. Three mirror systems were included in the package of instruments together with a collimated proportional counter for the higher energy X-rays up to about 12 keV. The mirror telescopes were able to position sources to better than one arc minute, a considerable improvement on the performance of the scanning instruments of the Uhuru satellite. Pointing of the Copernicus satellite could be maintained to about one arc second in a specified direction for about one orbit of the satellite (100 minutes). This remarkable accuracy was obtained with high-precision gyroscopes. In the observations of bright stars, the star itself could be used to provide a reference signal, and the accuracy of pointing could be maintained to 0.02 arc second. The superb performance of the Copernicus satellite allowed X-ray measurements to be made that did not need correction for errors in pointing: the beam widths of the X-ray telescopes were more than one hundred times greater than any such errors.

Shortly after the launch, the Copernicus X-ray instruments were used to observe the Cyg X-3 source. The ability to make pointed observations was immediately rewarded with the discovery that a regular modulation of 4.8 hours could be seen in the signals from the source. Scientists working with data from the Uhuru satellite also reported the same periodicity, but the evidence was obtained in a roundabout way that involved using a search technique in which trial periods are used to superimpose lengthy streams of data. A year later the Copernicus satellite was used to make simultaneous X-ray measurements of the source whilst infrared observations were being made by Neugebauer and his colleagues of the California Institute of Technology. A beautiful correlation of the 4.8-hour period was found in the infrared and X-ray light curves—there could be no doubt that the X-ray source had been identified with a counterpart.

The Cyg X-3 source, in spite of being identified as an X-ray, infrared and radio source, remains somewhat of a puzzle. Unfortunately there is no visible counterpart because of the large amounts of gaseous material and dust in the line of sight to the object. The absence of a visible star highlights the value of traditional astronomy; techniques developed over many years can be used to determine the nature, distance and motion of visible stars, all crucial parameters to the understanding of a binary system containing an X-ray source. In the case of Cyg X-3, the picture is thus incomplete, but the picture resulting from the studies of radio, infrared and X-ray measurements is of a double-star system probably consisting of a neutron star and a dwarf star in orbit about their common centre of gravity with a period of 4.8 hours.

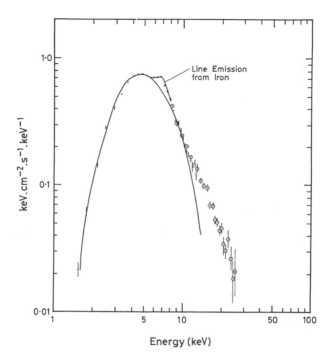

Fig. 5.3 The spectrum of the Cygnus X–3 X-ray source. This was the first spectrum obtained that had clear evidence of line emission. In this case the characteristic X-ray line from iron appears at about 7 keV

More insight into the nature of the Cyg X-3 system can be gained by making detailed studies of the X-ray emission and how it varies with the phase of the oribital period. Such measurements became possible with instruments that were carried on the eighth OSO and the Ariel V satellite. These instruments consist of proportional counters containing many cells separated by thin wires. Each cell is operated in anti-coincidence with its neighbours, so that a cosmic ray which triggers more than one cell is automatically rejected (see Fig. 2.3). This special detector, which was first introduced by scientists at Goddard Space Flight Center, gives a very low background noise and can thus be used to observe very weak sources. Special care is taken to ensure that the energy or spectral resolution is optimized.

When the Cyg X-3 source was observed with the Ariel V instrument the spectrum obtained contained a feature at about 7 keV which can be seen in Fig. 5.3. The observation of a feature in the spectrum of an X-ray source was an important discovery, not only as an aid in understanding the Cyg X-3 system but also because it raised the possibility of making studies of the spectral lines arising from the high-temperature plasmas that were believed to exist in many cosmic X-ray sources; for example, we have seen in Chapters 3 and 4 how detailed study of the emission lines in the solar-X-ray spectrum has provided a powerful technique for exploring the conditions in the active regions on the Sun.

In 1975 and 1976 the OSO-8 instrument made additional observations of the Cyg X-3 source which showed that the emission feature shifted slightly in energy during the 4.8-hour cycle of the binary system. It was not possible to determine exactly which ionization states were changing because the energy resolution of a proportional counter is not sufficient; however, it was clear that the emission feature was characteristic of iron atoms. The shift in energy of the emission feature is an important clue to understanding the system; previous measurements have been unable to detect any changes in the spectrum except for a modulation of intensity by a constant factor in every cycle.

Attempts to provide greatly increased spectral resolution with Bragg crystal spectrometers were made with instruments carried on the Dutch ANS satellite and the British Ariel V satellite. The weaker signals from cosmic X-ray sources were insufficient to give results with these instruments, which were no more than scaled-up versions of the conventional spectrometers that had produced detailed X-ray spectra of the emission from the Sun. We can expect similar results from cosmic X-ray sources when more sensitive spectrometers have been developed and flown. Promising results have already been obtained by Culhane and Zarnecki, who, using a Skylark rocket in 1976, reported the first detection of line emission with a crystal spectrometer—in this case from the supernova remnant Puppis A. An advanced instrument from Columbia University on the OSO-8 satellite was reported, in 1977, to have detected the iron line in the spectrum from the Cyg X-3 source.

We have described two characteristics—regular variability and spectra—which have become valuable indicators of the nature of galactic X-ray sources. Another feature, commonly found in galactic sources, is that they may only be observable for limited periods. The transient nature of such sources has been observed over a wide range of time scales. Some sources have been observed to increase to a peak intensity over a period of a few days; the flux then diminishes over an interval of several months, until it is no longer detectable. At the other extreme, transient phenomena have been observed from sources known as 'bursters', where the duration of the events is measured in seconds.

The first report of X-ray bursts in the 1–10 keV range were obtained from instruments on the ANS satellite. Two X-ray bursts were recorded from three independent detectors, which were pointed at the strong X-ray source 3U 1820–30. It was possible to determine their origin to a small region of uncertainty, which included the globular cluster NGC 6624. These observations were almost immediately confirmed by workers using the SAS-3 satellite. Many other examples of these bursting sources were subsequently discovered by the SAS-3 workers. By 1977 about twenty such sources had been discovered. Some of the sources display quasi-regularity in the occurrence of the bursts and one source, known as the 'Rapid Burster', has the remarkable characteristic of bursting in a regular fashion, then ceasing for a while after a particularly large burst. Optical bursts have also been discovered by Grindlay and co-workers, who in 1978 reported that X-ray and optical bursts had been detected simultaneously from the source known as MXB 1735-44. The burst profiles were qualitatively similar and were coincident to the accuracy of timing in the data from the SAS-3 satellite ($\pm$ 1 sec). These observations indicated that the burst of optical emission arose from a region very close to the X-ray source.

As the diverse characteristics of galactic X-ray sources came to be recognized, there was a natural tendency to ascribe a new class to each of these characteristics. However, by 1978 Margon had argued that if one considers the optical counterparts of the identified sources (apart from the supernovae remnants), then all but one or two sources would fall into two categories: those that shared the properties of the Sco X-1 source and those that were associated with very luminous stars. The varied spectra, slow pulsations, transient nature and bursting phenomena could be seen in examples from the two classes of optical counterparts. Margon restricted his consideration to the X-ray sources having luminosities in the range 10^{36}–10^{38} erg/sec. At these luminosities all sources within the Galaxy can easily be detected. It is now known that another class of low-luminosity sources also exist. The nearby star Capella is an example of this class as it has a luminosity 1,000 times less than the weakest sources in Margon's classification. These low-luminosity sources are probably members of a very large population, the more distant members of which can be observed only when instruments of much higher sensitivity become available.

5.3   X-rays from outside the Galaxy

Our Galaxy belongs to a group of more than twenty other galaxies; the nearest member is a cloud of stars known as the Large Magellanic Cloud which is at a distance of about 50 kpc. The Small Magellanic Cloud is also quite close at 60 kpc. These two clouds can be seen as patches of diffuse light and the nebulosity can be resolved into its constituent stars with the aid of a small telescope. Several sources have been detected in these clouds; a binary source in the Small Magellanic Cloud (SMC X-1) displays eclipses with a period of 3.9 days and its luminosity can reach the value of 3×10^{38} erg/sec. Another external galaxy can be seen from the northern hemisphere, at latitudes greater than 30° N, as a patch of light about a third of a degree across in the constellation of Andromeda. This galaxy, the Andromeda Nebula of M31, is similar to our Galaxy in its form and also has satellite galaxies similar to the Magellanic Clouds. If the Andromeda Nebula contains a distribution of X-ray sources similar to that of our Galaxy, then at its distance of 670 kpc we can predict the integrated emission in the X-ray region. It turns out that Andromeda is indeed an X-ray source, and the overall intensity is within a factor of 10 of the prediction. The X-ray emission from our Galaxy and the Andromeda Nebula is, however, almost insignificant when compared with the amounts of energy liberated in the X-ray region from other less normal galaxies.

The first extra-galactic source to be identified, M87, is at a distance of 13 megaparsecs (mpc), twenty times more distant than M31. The emission from this source amounts to 10^{43} erg/sec, which is almost 10,000 times the emission from M31. The galaxy M87 is a prominent member of the cluster of galaxies in the constellation of Virgo, which lies in the direction of the pole in our Galaxy. M87 is no ordinary galaxy as it has a jet of material which apparently originates in a region of violent activity within the galaxy. It was natural to assume that the X-ray emission would also arise from this active galaxy. Surprisingly, the data from the Uhuru satellite clearly indicated that the emission from that source came from a much larger region, more in keeping with the cluster of galaxies of which it was a member. Several other examples of emission from clusters of galaxies were also found by the Uhuru satellite. The clusters in the Coma, Perseus and Centaurus regions were all shown to have large amounts of X-ray emission and the study of these and many other examples has now reached an advanced state, as we shall see in Chapter 11.

The nature of the extra-galactic sources that were not associated with clusters became of great interest when it was discovered that the X-ray emission from one of them was measurably varying. When the measurements made with the Uhuru satellite on the Cen A source were compared with later measurements with the Copernicus satellite, there was strong evidence that the X-ray emission had varied between 1970 and 1973. Data from the OSO-7 satellite confirmed that the source was indeed varying in an interval as short as a few days. The history of the intensity-measurements of the source can be seen in Fig. 5.4.

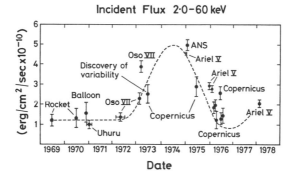

Fig. 5.4 The variability of the X-ray emission from the galaxy Centaurus A. The first evidence that galaxies could contain a compact, very energetic, X-ray source was obtained when this source was found to have varied by a substantial factor between measurements in 1970 by the Uhuru satellite and measurements in 1973 by the Copernicus satellite. Plate 10 shows a photograph of this galaxy

Cen A is the nearest of the galaxies characterized by their intense radio emission; for this reason they are called radio galaxies. The radio emission comes from the centre of the galaxy and from two enormous extended regions on opposite sides of the galaxy (see Plate 10). These two regions are believed to be the remnants of explosions which occurred several million years ago. The intense radio emission from the two lobes is an indication that very energetic particles are still present in these regions. Before the discovery of variability of the X-ray emission, it was thought that these energetic particles might also be responsible for the observed X-ray emission. However, the observation of variability ruled out this possibility, indicating instead that the X-ray emission must arise from the active centre of the galaxy. This has been confirmed in measurements made with the SAS-3 satellite, which located the sources of X-ray emission to within a few arc seconds of the central region of the galaxy.

The extra-galactic sources thus fall into two categories. Just as the sources in the Galaxy can be described as either diffuse or point sources, so the extra-galactic sources are diffuse in the case of the clusters of galaxies, or point-like in the case of active galaxies. Many other examples of these compact extra-galactic sources are now identified and we shall describe in chapter 10 how it has been possible to measure the spectra and variability of other examples.

The final example of X-ray emission from outside the solar system is the diffuse X-ray background that was discovered in the pioneering rocket flight of Giacconi and his colleagues in 1962. The nature of this background is the subject of intense study, both from a theoretical and observational point of view. The final chapter of this book will describe the status of these studies and their implications on our understanding of the very large-scale structure of the Universe.

6 · Supernovae and their remnants

6.1 Stellar evolution and supernovae

There are about one hundred thousand million (10^{11}) stars in our Galaxy. The vast majority of these, including the Sun, are extremely stable and maintain a remarkably constant output of radiation over millions of years. However, in contrast, novae and supernovae are stars that explode dramatically. Some novae may be X-ray emitters, and these will be discussed in Chapter 7. In this chapter we will deal only with supernovae.

We now know a supernova to be a violent end-point in the life of a star, so we will sketch briefly the stages in a star's evolution from birth to death. The nature of the life-cycle of a star is determined by two major influences. The first of these is the force of gravity. A star is born from a cloud of gas and dust which assembles in a galaxy and contracts as a result of the mutual gravitational attraction of its constituent particles. Once begun, the contraction or collapse of the cloud continues unchecked but during the process the temperature in the interior of the cloud steadily increases. Between 10^5 and 10^8 years later, depending on the initial mass of the gas cloud, the temperature becomes high enough for thermo-nuclear reactions to begin deep within the nascent star. The first reaction to start up causes the fusion of hydrogen to form helium nuclei in the so-called proton–proton cycle. When two light nuclei are fused together to form a heavier nucleus, it is found that the mass of the new nucleus is slightly less than the sum of the individual masses. This difference in mass appears in the form of energy according to Einstein's famous relation $E = \Delta mc^2$, where Δm is the mass destroyed, c is the speed of light and E is the quantity of energy released.

Once begun, these fusion reactions become the second major influence on the life-cycle of the star. The energy released in the reactions temporarily halts the collapse of the gas cloud because it is sufficient to balance the energy of gravitational collapse. Energy is transferred from the centre, or core, of a star—where the reactions are taking place—to the surface, where some of it is radiated away. Thus the collapsing cold-gas cloud has become a fully fledged star with a surface temperature which, depending on the initial mass, lies in the range 10^3–10^5 K.

This stable state of hydrogen fusion is found in the majority of the stars in the Galaxy. Stars in this situation, including our Sun, are said to be part of the main sequence. As we have seen above, stars remain fusing hydrogen for a lengthy period, which depends on their initial mass, but from this point on it is the balance between gravitational collapse and the release of nuclear energy that determines the future evolution of the star.

Although the nuclear-fuel reserves of a star are considerable, they are not unlimited. When the hydrogen in the core is used up, gravitational contraction begins again, since the energy release from hydrogen fusion has ceased. The energy of gravitational collapse raises the temperature of the core further, until it is high enough for the fusion of helium nuclei, which form carbon and oxygen. The energy released in these reactions again halts the gravitational collapse and the situation remains stable until all the helium is used up. Later stages of this process follow, involving the fusion of successively heavier elements up to iron. At this point in the sequence no further energy-releasing reactions are possible. In addition, because the abundances of the heavier elements are less than those of the lighter elements, the intervals during which collapse may be checked by nuclear-energy release grow shorter with each successive stage.

Before discussing the ultimate collapse of a star, which can lead to a supernova, it is worthwhile explaining a graphical presentation of stellar evolution that will also be of use in later chapters—we refer to the Hertzsprung–Russell or HR diagram which is shown in Fig. 6.1(a). In this diagram the luminosity or absolute magnitude of each star is plotted against the star's surface temperature. The resulting diagram is clearly far from random. The majority of the stars lie in the band labelled 'main sequence'. The position of the Sun is indicated. Stars, more massive than the Sun, use up their nuclear fuel more rapidly, have a correspondingly higher surface temperature and are more luminous. These stars populate the upper left-hand portion of the main sequence. Conversely, low-mass stars are found in the bottom right-hand corner of the diagram.

Superimposed on the HR diagram, in Fig. 6.1(b), is the schematic evolutionary track for a star with about the mass of the Sun. The gravitational contraction phase, the arrival of—and residence on— the main sequence and the progress to the red-giant part of the diagram are shown. The last stage takes place after the hydrogen in the core has been consumed. Thereafter the evolutionary track becomes uncertain. For stars of less than half the mass of the Sun, the fusion cannot proceed even as far as the helium stage. Such a low-mass star will soon cross the main sequence and end up in the white-dwarf region. These objects can collapse no further; they radiate away their residual energy and eventually disappear from view.

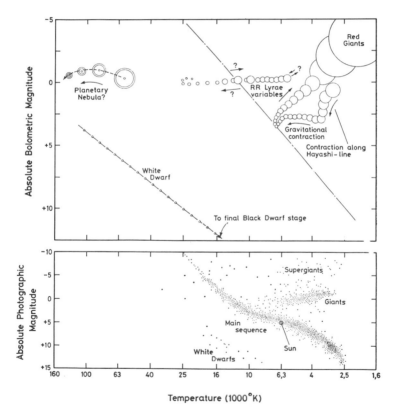

Fig. 6.1 (a) The Hertzsprung–Russell diagram or plot of stellar magnitude (luminosity) against surface temperature (b) A schematic evolutionary track of a solar-mass star and its relation to the main sequence. Bolometric magnitude refers to the total radiation output of the star

In the case of the star of about 1 M$_\odot$ (one solar mass), whose path is shown schematically, helium fusion can occur and thus the star can spend some time in the red-giant region. However, it can do no more than fuse helium and must then cross the main sequence with its nuclear energy spent, bound for eventual extinction at the bottom of the white-dwarf branch of the diagram. Along the way it may pass through a stage of variability as indicated.

From the point of view of the present chapter the stars of greater interest are those with more than four times the mass of the Sun. Such objects are capable of following the evolutionary trail to the point where no further energy may be extracted from fusion reactions. During these late stages of evolution, the nuclear fuel is used up very rapidly indeed. In this situation a final collapse can occur in a time measured in seconds. Such a sudden catastrophic collapse subjects the core of the star to enormous pressure. In this situation the electrons and protons in matter can no longer exist independently. Instead they are forced together to form neutrons. Thus a neutron residue of between 1 and 3 M$_\odot$ and of incredibly high density is created. Such objects are called neutron stars and they will be the subject of a more detailed discussion in a later chapter. For even more massive stars, the pressure caused at collapse is sufficiently high to drive the residue beyond even nuclear densities. In this circumstance a black hole is thought to be created and these esoteric objects will also be the subject of later discussion.

The sudden prevention of further collapse in the core leads to enormous pressures and temperatures. The remainder of the star, which is collapsing behind the core, will suddenly experience the high pressure and so the implosion or collapse is converted by a bounce into an enormous explosion leading to the violent ejection of a substantial fraction of the mass of the star. This material is ejected into the interstellar medium at velocities in excess of 10^4 km/sec leaving behind the small dense core. Both of these remnants of the catastrophic collapse are observable in a number of different spectral ranges, among which the X-ray region is of particular importance.

6.2 The condensed remnants of supernova explosions

Following the collapse of the stellar core, the most probable central remnant is a dense neutron star, a body with a mass similar to that of the Sun, but having a diameter of only 10 km. Such an object is unbelievably dense; a matchbox filled with neutron-star material would weigh 10^{12} kg at the Earth's surface. The first observational evidence for the existence of such objects came with the discovery of radio pulsars by Hewish and Bell and their collaborators who were operating a new radio telescope at Cambridge in an investigation of quasars (see Chapter 10). Weak radio signals of variable intensity were found to originate from a point in space. Although these signals disappeared from time to time, they were, when detectable, in the form of a succession of pulses of amazingly constant period. These signals were clearly identified as being of celestial origin when radio pulse trains from three further directions in space were discovered by the Cambridge group. At the time of writing well over three hundred of these objects have been detected and catalogued.

Many theories were advanced to explain the new phenomenon. These included vibrating white dwarfs or neutron stars, very close binary systems, a planet in orbit around a neutron star and many others. All of these were eventually discarded in favour of an explanation based on rotating neutron stars, which was first proposed by Thomas Gold.

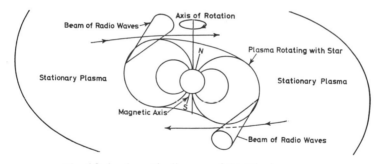

Fig. 6.2 A schematic diagram of the spinning nuetron-star model of pulsars as proposed by Gold

The main features of a current version of this model are illustrated in Fig. 6.2. Ordinary stars rotate slowly (the Sun's rotation period is about twenty-seven days) and have weak magnetic fields (about 1 Gauss for the Sun). In the sudden collapse of a star, which gives rise to a supernova explosion, the condensed core has a diameter of only 10 km compared with the much larger value before its collapse. But stellar angular momentum must be conserved. Hence, if the star's radius is drastically reduced, its rotation rate must be greatly increased in order to maintain the angular momentum. It was estimated by Gold that the collapse of a typical star would leave the resulting neutron star rotating at a rate of about once per second or faster. The radio pulsar periods were all in this range. The star's magnetic field is also affected by the collapse. Magnetic energy, like angular momentum, must be conserved through the collapse. Since the volume available for the magnetic field is negligibly small compared with the pre-collapse value, the field strength is increased enormously. Gold again estimated that a neutron star's magnetic field should be in the neighbourhood of 10^{12} Gauss. The existence of such a strong magnetic field ensures that any plasma surrounding the star is tightly held by the field and so rotates with the star. As the radial distance from the surface of the star increases, the angular velocity of the rotating plasma must likewise increase until it is rotating with the star at almost the speed of light. Plasma is then ejected at high velocity into the surrounding region but, before leaving the high-field region, the plasma electrons radiate radio-frequency pulses by the synchrotron process. It is these radio pulses, received on Earth, that give rise to the pulsar phenomenon.

Although the above model has been considerably refined since it was first proposed, it remains in essence the basis of our understanding of pulsars. The correlation between the occurrence of a supernova and the existence of a pulsar is not one for one, and this has lead to some difficulty, but the ability of a supernova collapse to create a neutron star is not in doubt. A more detailed discussion of the nature of neutron stars will be given in Chapter 9, where the emission of X-rays from binary-star pairs will be discussed.

6.3 The Crab Nebula

In the year AD 1054 Chinese astronomers chronicled the appearance of a new star, or guest star, in the heavens. For about a month it was visible even in daylight and was clearly a most unusual object. We now believe that the object seen by the Chinese was a supernova explosion. Its extended remnant, which remains visible today, is called the Crab Nebula (see Plate 11).

This object provided a major puzzle for astronomers for many years. Its emission stretches from the radio range through to gamma-ray energies. The nebula shows evidence of continued activity even up to the present time. Wisps and filaments of gas are observed, in the central regions of the source, to undergo rapid changes. The emission spectrum is continuous throughout the entire range of observation and from its polarized nature appears to be due to synchrotron radiation from high-energy electrons in an extended magnetic field. Although the supernova explosion occurred almost 1,000 years ago, the nebula is still radiating 10,000 times more energy than the Sun and it was this fact, more than any other, that appeared to defy explanation until 1968 when a pulsar was discovered in the centre of the nebula.

The Crab pulsar has remained unique since its discovery. First, with a pulse period of 0.0331 sec, it is the fastest pulsar known. Following the explanation of the pulsar phenomenon given in the previous section, we can say that the 'clock' for this rotation period is a rapidly rotating neutron star of approximately 1 M$_{\odot}$. Second,

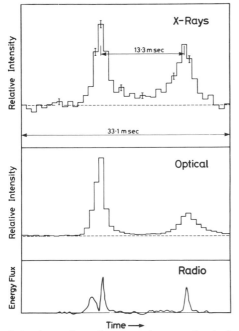

Fig. 6.3 The pulse profiles from the neutron star in the Crab Nebula at radio, optical and X-ray wavelengths

it is the only pulsar to give a pulsed signature in all spectral ranges in which it has been studied. Pulse profiles for the radio, optical, X-ray and gamma-ray ranges are shown in Fig. 6.3. Detailed measurements of the period of the pulsar show that the neutron star's rotation is gradually slowing down at a rate of 3.5×10^{-8} sec per revolution per day. This is a most important observation. Using this value for the decrease in period we can calculate the rate at which the spinning neutron star is losing its rotational energy. This rate of energy loss turns out to be essentially equal to energy radiated from the nebula. Thus, the rotating neutron star is supplying power to the entire nebula. It does this, in the main, by injecting high-energy accelerated electrons into the nebula where they radiate by the synchrotron process as described above.

X-ray observations such as the power-law nature of the X-ray spectrum and the discovery, in 1972 by Novick and his colleagues at Columbia University, that the X-ray emission was polarized are playing a major role in helping us understand the details of the nebula-radiation mechanism. Both observations confirm the prediction based on radio and optical data that all of the emission from the nebula is due to the synchrotron process. However, there are a number of further questions that arise. For example, it is not at all clear how the accelerated electrons leave the neighbourhood of the pulsar and begin to interact with the more extended magnetic field of the nebula. This question might be resolved if one could obtain an X-ray map of the Crab and thus determine how the energetic electrons are distributed within the nebula. This is a difficult problem since the diameter of the nebula as seen from the solar system is quite small—between 1 and 2 minutes of arc. Note that, for comparison, the apparent diameter of the Moon is 30 minutes of arc. In order to obtain useful data on the distribution of electrons, an X-ray map of a few arc second resolution is needed. Prior to the flight of the HEAO-2 (Einstein) spacecraft, X-ray telescopes of this quality were not available.

It happens, however, that the Crab Nebula lies in the path of the Moon and about once every ten years a number of lunar occultations or eclipses of the nebula occur. As the edge of the Moon passes in front of the nebula it hides (or exposes) progressively more of the X-ray emission. From a detailed knowledge of the Moon's motion and position, it is possible to calculate the location and size of the region emitting X-rays and the distribution of X-ray intensity throughout this region. This kind of observation, much used in the early days of radio astronomy, was first carried out on the Crab's X-ray emission in 1964 by Friedman and his colleagues at the NRL (see Chapter 1). It was repeated ten years later by a number of groups using more sensitive equipment. The solid circle superimposed on the nebula in Plate 11 indicates the volume of the nebula that emits X-rays. Lunar-limb positions are also shown, as an aid to appreciating how an occultation observation is conducted. The X-rays appear to be uniformly produced and radiated within this spherical volume. However, it is interesting to note that the position of the pulsar, which is indicated with an arrow, does not coincide with the centre of the X-ray emission, indicated by a cross.

Seward and his colleagues at the Lawrence Livermore Laboratory in California, who undertook one of the 1974 lunar-occultation observations with a sounding rocket, have pointed out that the off-centre location of the pulsar and the relatively uniform brightness of the nebula in X-rays both indicate that the electrons which give rise to the X-ray emission by the synchrotron process must undergo acceleration throughout the nebula and not just in the immediate neighbourhood of the pulsar. Some time before this observation, Scargle and Barnes had proposed that magneto-hydrodynamic waves originating at the pulsar were propagated throughout the nebula. These waves could explain the observed activity in the optical filaments and could also accelerate the electrons. Hence, although the basic role of the central neutron star as the power-house of the Crab Nebula remains unchallenged, the X-ray observations demonstrate the need for a more detailed investigation of how the energy is transferred from the pulsar.

Seward's lunar-occultation observation drew attention to one further interesting feature of the Crab's X-ray emission. Even after the nebula was completely covered by the Moon, a significant low-energy (0.5–1.0 keV) X-ray flux was observed by the rocket-borne detectors. Although this flux is only 7 per cent of the intensity of the corresponding signal from the main nebula, its existence has been confirmed by Charles and Culhane using the Mullard Space Science Laboratory's X-ray telescopes on the Copernicus satellite. Two regions of low-energy emission, beyond the boundary of the main Crab X-ray source, were found by the telescopes and these regions together provide the signal observed by Seward's instrument. These lower-energy X-rays have a different spectrum from that of the nebula and are consistent with being generated in a hot plasma at a temperature of around 7×10^6 K. We will see in the next section that the existence of such hot plasma adjacent to the Crab Nebula is not unexpected.

6.4   The extended remnants of supernovae

The Crab Nebula is almost unique among galactic remnants in possessing a known pulsar that is obviously still supplying power. The majority of extended supernova remnants have somewhat different properties from those of the Crab. There are about a hundred known supernova remnants in our Galaxy. In the main these have been identified by the presence of a non-thermal radio source, whose emission is polarized. This emission is thought to be due to the synchrotron process, but it is usually less bright than that of the Crab and we will see later that the electrons interact with a magnetic field in a rather different way. Usually, the radio emission is concentrated in a ring—or part of a ring—structure. Optical filaments are also observed (see Plate 12). These are generally concentrated in and around the radio ring and they emit a line spectrum in visible radiation Measurements of Doppler effects on individual lines indicate that the outer shell of the remnant, as delineated by the radio maps, is expanding radially. In cases where the appropriate radiation can be transmitted through the interstellar medium, supernova remnants are found to be extended X-ray sources. Finally, the identification of a remnant may be secured by establishing a relationship with a recorded supernova event. The Crab Nebula presents the best known case but there are many other examples.

In order to understand the different features of a supernova remnant it is necessary to examine what happens in the region around the exploding star. We stated earlier that after a neutron star is formed the outer layers of the star rebound from its dense surface and are ejected violently into interstellar space. This mass of ejected material is substantial: for a massive star it could be as high as ten times the solar mass. However the interstellar medium is not empty; the gas and dust from which stars are formed exists there with a density of between 0.1 and 1 particle per cm^3, and gas clouds with much larger densities than this are also known to exist in the interstellar medium. The ejected material moves away from the site of the explosion in the form of an expanding sphere and at a speed of as much as 15,000 km/sec. For the first ten years after the explosion the sphere of material expands freely through the interstellar medium. Although it sweeps up the interstellar gas and dust into a spherical shell, there is initially very little matter swept up compared with the large mass of ejected material. However the very high velocity of the ejecta produces a shockwave, formed at the leading edge of the swept-up interstellar gas, which is thereby raised to a very high temperature. The temperature depends strongly on the expansion velocity and will be around 10^9 K for a velocity of 15,000 km/sec. However, this hot, swept-up gas makes little or no contribution to the X-ray output of the supernova because the swept-up mass is still a tiny fraction of the ejected mass.

About one hundred years after the explosion the situation has changed significantly. The amount of swept-up material has become comparable with the mass of the material originally ejected. However, the act of sweeping up this material has decelerated the shock-wave to a velocity of around 5,000 km/sec. This reduction in velocity means that the swept-up material is heated to a temperature of less than 10^8 K but its mass is now large enough for a substantial X-ray flux to be detected from the remnant. These are X-rays from hot gas like those generated in the solar corona (see Chapter 3). Thus the X-ray emission from this shock-heated gas occurs without any power being supplied by a pulsar, even if one should still exist at the site of the explosion. The situation is complicated by the piling-up of the innermost shells of ejected material against the slowly moving outer layers. This leads to the creation of a lower velocity shock-wave and to the production of gas at a lower temperature than that of the interstellar material which has been heated by the initial shock. We will discuss two examples of this behaviour in the next section.

Returning to the progress of the primary shockwave, it continues to plough through the interstellar gas and to sweep up more and more material, so that after 1,000 years the mass of the swept-up shell is considerably larger than that of the ejected material. The temperature at the shockwave also continues to fall since the shock velocity is continually declining. When the latter has reached a value of around 100 km/sec, the shock-heated gas temperature will have fallen to around 10^6 K. At this temperature the gas can cool rapidly because it is able to radiate more efficiently than at higher temperatures due to the many emission lines in the spectrum. The shock velocity continues to fall until it reaches a value comparable to that of the interstellar gas, that is around 10 km/sec. At this point the remnant has ceased to exist. This stage is reached approximately 100,000 years after the original explosion.

The progress of the shock through the interstellar medium leads to two other observable effects, in addition to the creation of a hot X-ray-emitting shell. Besides the interstellar gas, the interstellar magnetic field is swept up by the shock and concentrated in a shell. High-energy cosmic-ray electrons, which exist throughout the interstellar medium, become trapped in these concentrated magnetic-field structures where they emit radio waves by the synchrotron process. These fields are, however, too weak to permit the generation of synchrotron X-rays. The only source of X-rays is thus interstellar gas heated by shockwaves.

The interstellar gas is not generally uniform. It includes many small high-density regions and when the advancing shockwave from a supernova explosion heats these regions, they cool much more rapidly than the rest of the interstellar medium, because of their high density. Thus these areas appear as cool filaments embedded in the hot X-ray-emitting gas and, because their temperature is low, they radiate visible light. The Cygnus Loop (Plate 12) also provides a very beautiful example of a remnant with optical filaments. This object is also known as the Veil Nebula.

In the next two sections we will discuss some examples of young and old supernova remnants in the light of the above description of how a shockwave is formed after the explosion. However, the description presented here should make it clear why the Copernicus observations of the Crab Nebula showed evidence for the existence of regions of hot gas outside regions powered by the pulsar. These faint regions of low-temperature X-ray emission are located just where we would have expected the expanding shockwave to have reached 1,000 years after the supernova explosion.

6.5   Cassiopeia A and Tycho's Star—a pair of young supernova remnants

The Crab Nebula is unique because of its pulsar. It is also a bright X-ray source with a strength of around 1,000 Uhuru sky survey counts per second in the 2–10 keV X-ray range. Recall, however, that it has associated faint-X-ray emission of about 80 Uhuru counts per second, which we believe originates from shock-heated interstellar gas with a temperature of around 7×10^6 K.

Even before their identification as X-ray sources, radio astronomers had found no evidence for the presence of pulsars in either Cas-A (Cassiopeia A) or Tycho. However, both objects have a shell-like radio appearance, which exhibits a complex filamentary structure when examined with high-resolution radio telescopes. Both remnants are younger than the Crab Nebula with ages of 300 and 400 years respectively. Tycho's star was seen to explode in 1572 but the Cas-A supernova was not observed on Earth, although its present radio size and expansion velocity suggest that the explosion should have occurred at the beginning of the eighteenth century. However, the source is located in a relatively obscured region of the Galaxy. Although at comparable distances to that of the Crab, Cas-A and Tycho have X-ray fluxes of 50 and 10 Uhuru counts per second respectively. These low levels of flux again appear to argue against the presence of a pulsar as the main power source.

Using the Copernicus satellite, Charles and Culhane were able to map Cas-A in the energy range 0.5–3.0 keV with an angular resolution of about 2 arc minutes. Since the diameter of the radio source is only 6 arc minutes, the resulting X-ray map was necessarily rather crude (see Plate 13). In particular the resolution is not adequate to demonstrate the existence of the shell-like structure that is apparent in the radio map. However, low-resolution radio maps, made several years ago, are very similar in appearance to the X-ray map. Hence, it seems likely that a shell structure will become visible when higher resolution X-ray maps are available. The map in Plate 13

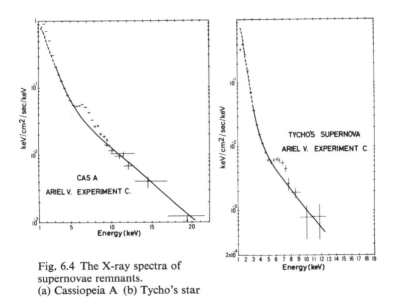

Fig. 6.4 The X-ray spectra of supernovae remnants.
(a) Cassiopeia A (b) Tycho's star

shows no evidence for the presence of an active pulsar at the centre of the remnant. This, coupled with the low X-ray brightness of both Cas-A and Tycho relative to the Crab, strongly suggests that shock-heated interstellar gas is responsible for the observed X-ray emission.

A more detailed examination of the X-ray spectra of these objects has provided further evidence for the X-rays arising from a hot gas. This evidence was obtained for both sources by Davison, Culhane and their co-workers using the Mullard Space Science Laboratory's proportional counter spectrometer on the Ariel V satellite (see

Plate 5). These two spectra are shown in Fig. 6.4. Both show a continuous form, though with a marked change of slope at around a photon energy of 5 keV. However, the most significant feature is the existence of a line at an energy of 6.7 keV in the spectra of both sources. We have seen that the hot plasma in the solar corona (Chapter 3) has a spectrum with many emission lines. The feature at 6.7 keV is mainly due to the strongest or resonance line of the Fe XXV helium-like ion (the atom has lost all but two of its electrons, see Section 3.1). Other lines are now known to exist at lower energies. In particular, lines due to silicon ions have also been found with proportional counters. The presence of these lines makes it quite clear that the X-radiation from these two young supernova remnants comes from hot plasma and not from electrons radiating in magnetic fields by the synchrotron process.

The change in slope evident at 5 keV arises because of the existence of gas with at least two distinct temperatures in the remnants. These temperatures are indicated in Fig. 6.4. At ages of 300 and 400 years respectively, Cas-A and Tycho have probably just reached the stage in their evolution where the mass of swept-up interstellar medium material is equal to the mass of originally ejected material. In these circumstances an additional lower velocity shockwave can propagate through the ejected matter, heating it to a lower temperature than that associated with the primary shock. Thus the higher temperature gas is responsible, in both objects, for the Fe XXV line emission while the lower temperature material emits the silicon lines.

Detailed studies of the X-ray spectra of both of these objects, using high-resolution Bragg spectrometers, will be of the greatest importance, as will the making of high-resolution X-ray maps. The latter can show the existence of clouds and other density irregularities in the interstellar medium, while the former will allow measurements of element abundances in both the ejected material and the interstellar gas. In the late stages of a supernova collapse the creation-rate for heavy elements may be much enhanced and the measurement of their abundances assumes considerable importance.

6.6 Old supernova remnants

The two young remnants that we have discussed are both less than 500 years old. Since they have only been expanding for this time, their diameters are less than 20 light years (6 pc) and the temperature of the associated shock-heated plasma is high. Although the distinction is necessarily somewhat arbitrary, we will class as old supernova remnants those objects having ages greater than 5,000 years and diameters greater than 50 light years (15pc). Since such objects have been expanding for longer periods, their shockwave velocities will be lower, and their temperatures correspondingly less than those we observe in Cas-A and Tycho.

A prime example of an old remnant, the Cygnus Loop, is illustrated in Plate 12. It is about 20,000 years old and has a diameter of 120 light years. For this object the temperature of the hot interstellar plasma is about 3.0×10^6 K. While X-ray line emission has not yet been detected, the existence of hot gas in this remnant has been confirmed by Woodgate and his co-workers. Using a narrowband filter, on a ground-based optical telescope, they detected the presence of Fe X and Fe XIV red and green lines in the spectrum of the supernova remnant. The reader will recall that the high temperature of the solar corona was conclusively established by the identification of these same lines in the coronal spectrum (see Chapter 3). An X-ray map of the Cygnus Loop shows a shell-like structure similar to that of the optical in Plate 12. However, the brightness of the X-ray shell is not uniform. There are large gaps in it, particularly in the southern part of the remnant. The bright regions probably indicate the presence of density gradients in the interstellar medium, which are in turn due to the presence of interstellar clouds. To the south the shock has probably encountered regions of low and uniform interstellar density which have insufficient material to radiate detectable X-ray fluxes.

The remnant located in the constellation of Vela (Vela X) is a very similar object. It is 15,000 years old, has a diameter of about 90 light years, and a hot plasma temperature very similar to that of the Cygnus Loop. X-ray maps also show a broken shell structure, which is once again probably due to the presence of interstellar clouds and gas-density gradients. This remnant is unusual in that it contains a pulsar within its boundaries. The Vela pulsar with a period of 0.07 sec is the most rapid of these objects known, apart from the Crab Nebula. The rate of increase of this period is consistent with the age of the remnant, deduced from its current expansion velocity and size. Hence the Vela pulsar is almost certainly the condensed remnant of a supernova explosion whose shockwave gave rise to the Vela X remnant. There is also an extended X-ray-emission region in the area immediately around the pulsar, but it is much fainter than the Crab Nebula (10 Uhuru counts per second as against 1,000 Uhuru counts per second). This emission is probably from the synchrotron process; its X-ray spectrum is quite different from that of the rest of the Vela remnant. Thus, in the case of this object, shock-heated plasma and synchrotron radiation are observable.

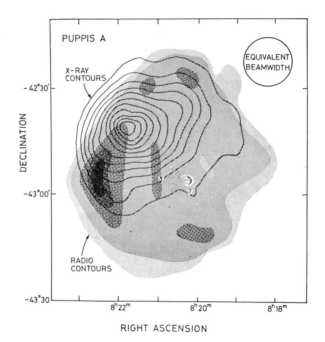

Fig. 6.5 Distribution of radio and X-ray emission from the Puppis-A supernova remnant. The hatched regions represent differing radio intensity while the contour lines indicate the distribution of 0.5–1.5 keV X-rays

Puppis A is a supernova remnant that has been studied in con-siderable detail. X-ray and radio maps of this object are shown in Fig. 6.5. The X-ray data were obtained with the Mullard Space Science Laboratory telescopes on the Copernicus satellite. The brightest part of the X-ray map does not coincide with the bright radio-emitting regions. Although most of the X-ray emission arises in a plasma at a temperature of 1.0×10^7 K, the bright area has a gas temperature closer to 4×10^6 K. This has been confirmed by the detection of X-ray line emission by Zarnecki and co-workers, who flew a Bragg crystal spectrometer on a Skylark sounding rocket (see

Plate 4). The line was from O VIII ions, which could not exist at a temperature as high as 1.0×10^7 K. Woodgate, Zarnecki and others have further strengthened the identification of cooler plasma by detecting the visible green line of Fe XIV in observations with ground-based telescopes. The cooler bright region is again probably due to the interaction of the expanding shockwave with an interstellar cloud.

Thus, in a few short years, X-ray observations have transformed our understanding of supernovae. We now clearly appreciate the distinction between the synchrotron-emitting regions, which immediately surround active pulsars (e.g. Crab and Vela), and the high-temperature interstellar plasma that is heated by the expanding shockwave from the site of the original explosion. We can expect further X-ray studies to reveal a great deal about the creation of elements in supernovae and the detailed structure of the interstellar medium.

7 · X-rays from binary stars

7.1 The dynamics of binary stars

Galileo's observations of the planet Jupiter, and the discovery of its moons, provided the first evidence that the Earth is not the centre of the Universe about which all the astronomical bodies move. Jupiter and its moons are an example of a gravitationally bound system where the moons are clearly in orbit about the planet. The Copernican view (the Earth and the planets are in orbit about the Sun) had to be accepted, with the support of this evidence that orbiting systems exist in which the Earth plays no part. From this direct observation of an orbiting system, the subject grew to become one of the most important branches of observational astronomy.

In 1803 F. W. Herschel discovered that Castor, the brightest star in the constellation of Gemini, is a visual double star whose components move around each other under the influence of their mutual gravitational attraction. The observation of double stars enables one to determine the masses of both components in some cases, and qualitative statements can usually be made about the masses even if only one of the components can be viewed directly. The masses of stars are only detectable through their gravitational interactions so this field of astronomy has a fundamental significance for astrophysics and especially for the study of those X-ray sources that exist in binary systems.

The true nature of orbital motion was first described by Kepler early in the seventeenth century. His three laws of motion for the planets about the Sun are:

1   Each planet moves in an orbit that is an ellipse, with the Sun at one of the foci (the other focus is empty).

2   The speed of the moving planet changes with distance from the Sun, so that the radius vector sweeps equal areas in equal intervals of time.

3   The squares of the times that it takes the planet to go round the Sun (the periods of revolution) are proportional to the cubes of the mean distances from the Sun.

The 'mean distance' in the third law is half the major axis of the ellipse. Kepler announced this third law in 1619, ten years after he had stated the first two.

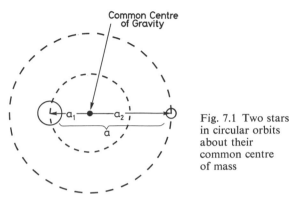

Fig. 7.1 Two stars in circular orbits about their common centre of mass

Although Kepler's laws referred to the motions of the planets about the Sun, the laws are also valid for the motions of double stars. Fig. 7.1 shows two stars orbiting in the special case of circular orbits. By careful measurement one can obtain the semi-major axis of the relative orbit of a star in a visual binary and from Kepler's third law one can then obtain the combined mass of the two stars as:

$$M_1 + M_2 = a^3 \big/ \theta^3 p^2$$

Where θ is the parallax of the star in seconds of arc, P is the period of the orbit in years, a is the semi-major axis in seconds of arc and $M_1 + M_2$ is equal to the combined mass of the two stars in terms of solar masses.

If one can also measure the absolute motion of both stars (i.e. relative to the background stars) it is possible to obtain the semi-major axes of both components of the visual binary which move in elliptical orbits about the mass centre of the two stars. The ratio of the masses of the two stars is then given by the ratios of the semi-major axes of the two orbits, that is:

$$a_1 : a_2 = M_1 : M_2$$

and

$$a = a_1 + a_2$$

so that it becomes possible to calculate the individual masses of the components M_1 and M_2.

For some binary stars, the fainter companion is not visible, but its presence can be inferred from the motion of the brighter component about the mass centre. If a_1 is the semi-major axis of this orbit, and since $a_1/a = M_2/(M_1 + M_2)$ we obtain:

$$(M_1 + M_2) \left(\frac{M_2}{M_1 + M_2} \right)^3 = a_1^3 \Big/ \theta^3 P^2$$

The German astronomer F. W. Bessel discovered that the bright star Sirius has motion which could be accounted for by the presence of an invisible companion. The companion was found eighteen years later, in 1862, by the American optician Alvan Clark while testing an eighteen-inch objective, the largest astronomical lens in existence at that time. His instrument disclosed a tiny satellite star (10,000 times fainter than Sirius itself) in the exact position predicted. When the usual relationship between the mass of a star and its luminosity was well established, it became clear that the companion to Sirius is a rather special star. At almost the same mass as the Sun, and with a normal surface temperature, the luminosity was much too low. In 1923 F. Bottlinger concluded that 'here we have to do with something entirely new', namely a *white dwarf* star.

In recent times dark companions have been found for certain nearer stars. In particular, Barnard's star, our second nearest neighbour, possesses a companion of only about $0.0015 \, M_\odot$, or about 1.6 times the mass of Jupiter. We can thus observe a second planetary system within a distance of 1.84 pc.

Many double stars are too closely spaced to be resolved as separate stars, even with the largest telescope. However, their double nature becomes apparent when spectral lines are observed and for this reason they are called *spectroscopic binaries*. The star Mizar was the first to be recognized as such a double system. In 1889, E. C. Pickering observed that the spectral lines of this star occasionally appeared double, and at other times single. On further study the changes were found to be periodic, the lines doubling at intervals of about ten days. He interpreted the changes as the result of orbital motion, which introduced Doppler shifts to the lines from the two stars. When the lines appeared double, one star was receding and the other approaching. Since the stars had identical brightness and spectra, the doubling occurred twice in each orbital revolution. The star

Mizar is thus a spectroscopic binary with a period of twenty and a half days.

Mizar consists of two similar A2* stars in which the components are both visible. In other binary stars only one of the components is visible. If we plot the magnitude of the shift in the spectral lines with time, we obtain a curve known as the velocity curve. We can calculate from this curve, not the semi-major axis itself, but the quantity $a_1 \sin i$, the inclination angle (usually given the symbol i) of a system, which is the angle the orbital plane makes with the normal to the line of sight. In this case we obtain from Kepler's third law:

$$\frac{(a_1 \sin i)^3}{p^2} = (M_1 + M_2) \left(\frac{M_2}{M_1 + M_2}\right)^3 \sin^3 i = \left(\frac{M_2^3 \sin^3 i}{(M_1 + M_2)^2}\right)$$

The last quantity is called the *mass-function*. If the second spectrum is visible we obtain a further mass-function of the form given above with M_1 interchanged with M_2. We can eliminate the inclination and obtain the ratio of the masses of the two components of the binary.

If the inclination of the orbital plane of a spectroscopic binary is 90°, then eclipses occur and we observe a variable star known as an *eclipsing binary*. The first example was discovered by J. Goodricke in 1782, who interpreted the variability of Algol in this way.

Eclipsing binary stars are rewarding to study as it is possible to determine the absolute dimensions of the system, providing one can also measure the velocity curves. It then becomes possible to obtain the masses of the two stars and their radii, which leads immediately to the density of the stars. H. N. Russell and H. Shapley developed precise mathematical methods, in the early twentieth century, for the analysis of such stars—with the result that stars in eclipsing binary systems are the most accurately known. O. Struve later showed, through detailed analysis of the spectra, that in close pairs the two components are in a state of interaction. Common gaseous envelopes and gas streams flowing from one star to the other have a very great influence on the evolution of such systems.

7.2 X-rays from binary stars

The first clear evidence that X-ray sources could exist in binary systems was obtained with the Uhuru satellite. Regular eclipses were seen in four X-ray sources (Her X-1, Cen X-3, 3U 0900–40 and

* A2 refers to the astronomers' classification scheme based on temperature.

3U 1700–37). Prior to these discoveries there had been suggestions on theoretical grounds that binary systems might provide the necessary conditions for the production of large amounts of energy, radiated in the X-ray region of the spectrum. The idea results from the simple consideration that material which falls on to a collapsed star can gain high velocities and this kinetic energy will be converted into heat when it impacts the surface of the collapsed star.

Other, more complex models have been put forward for the production of significant X-ray emission in binary systems where both components are normal stars. One of these possibilities was suggested by J. Bachall and others in 1973. Their model assumes that the binary consists of two ordinary—but magnetic—stars, which are linked by magnetic flux. If the stars are not rotating at the same rate as the orbital period, then the lines of force become twisted-up by the lack of synchronism and regions of very high magnetic flux are produced. Instabilities set in, similar to those which are thought to be responsible for solar flares. The result is the release of energy in the form of X-rays. The energy liberated, according to the model, is similar to that which is observed in X-ray sources. The other prediction, of rapid variations in the flux, is also observed in some X-ray sources. The mechanism is not, however, thought to be a common one in the known X-ray sources for two reasons: first, in the identified sources, there is no strong evidence for the existence of the required magnetic fields; and second, those stars which have the necessary magnetic conditions are not strong X-ray sources.

Another mechanism for the production of X-rays in a binary system, again with normal stars, was evaluated by Cooke, Fabian and Pringle in 1978. This process depends on the collision between the winds that are blown from the stars in the binary. Our Sun has such a stellar wind, which gives rise to the trapped particles of the radiation belts where the wind encounters the Earth's magnetic field. There is even occasional production of X-rays in parts of this outer atmosphere of the Earth, which is known as the magnetosphere. The production of energy, predicted in the model for interactions of stellar winds, can reach 10^{33} erg/sec in extreme cases. Such a flux will easily be detectable from nearby stars with instruments that are likely to be available in the 1980s. The strong X-ray sources listed in the early catalogues cannot be energized by this process as the X-ray luminosities lie in the range 10^{36}–10^{38} eg/sec. The model is thus inadequate for these sources. The accretion of material on to a compact star and the associated heating of this in-falling material remains the most likely mechanism for the production of intense X-ray emission in the binary sources.

Support for the accretion model was initially derived from the earlier observations of dwarf novae binary systems. It is known that old novae and stars called U Geminorum are binaries in a state of mass transfer, where one star—usually a main-sequence star—is losing mass, some of which falls on to the companion. This companion, which is probably a white dwarf, has anomalously high visible luminosity caused by the heating produced in the accretion of matter on the surface of the collapsed star. Before we describe the details of the mass transfer and phenomena of the accretion process we will discuss the evolution of stars in binary systems. The normal life-cycle of stars becomes significantly modified when there is mass transfer between the two stars.

7.3 The evolution of double stars

We have mentioned earlier that the observations of eclipsing binaries provide the most accurate information on the masses, radii and hence the density of stars. These systems can be used to test the theories on the evolution of stars. In the studies of the 145 brighter (greater than 8.5 magnitude) eclipsing binaries it is found that two groups are most prominent. The first group contains fifty-nine binaries where both components lie on the main sequence of evolution (see Chapter 6). The interpretation of their evolution presents no difficulty. Both components are likely to be of the same age and chemical composition. Evolution along the main sequence is very slow so that many such systems would be expected. The next most prominent group, with fifty-two members, are eclipsing binaries with characteristics similar to Algol (β Persei, the second brightest star in the constellation of Perseus). In this group the primary component is more massive, brighter and lies on the main sequence of evolution. The companion has considerably smaller mass but its optical spectrum and luminosity indicate that it has evolved to a point well above the main sequence.

A fundamental result of the theories on the evolution of stars is that it proceeds more rapidly with increasing mass. The time spent on the main sequence varies inversely as the third power of the mass:

$$T \propto 1 \Big/ M^3$$

In the Algol system the primary is approximately five solar masses (5 M$_\odot$) and the secondary one solar mass (1 M$_\odot$). The ratio of times spent on the main sequence should therefore be 1 : 125. In Algol the less massive star has evolved to a point somewhere between a subgiant and a giant. In the normal course of events this would have taken about 10^{10} years. This is clearly impossible since the more massive star would have evolved to a super-giant in less than 10^8 years, whereas it is still on the main sequence of evolution. This *evolutionary paradox* is the characteristic of eclipsing binaries in the group known as *Algol-like systems*. The clue to understanding the paradox came when it was suggested, by Crawford in 1955, that large-scale transfer of mass had occurred in these systems. The more evolved star had originally been more massive but a considerable proportion of its mass had been transferred to its companion. For Algol, 70 per cent of the original mass would have been transferred. It was not easy to accept this model of massive mass transfer until detailed calculations were presented by Morton in 1960. He showed that the internal structure of stars could survive in the transfer of large quantities of mass.

Morton's work indicated that the transfer of mass could occur on a relatively rapid time scale (compared with the normal evolutionary phases). The transfer begins when the originally heavier star expands, during normal evolution, to reach a critical radius. At this point the star fills a volume known as the *Roche lobe*. The flow of matter which then occurs is an important factor in the evolution of the binary star. It is particularly significant when the star has reached the stage of X-ray emission.

The Roche lobes associated with stars in close binaries can be described with the aid of Fig. 7.2. This shows the two stars and the contours of gravitational potential. The centre of the co-ordinate system is taken as the centre of mass of the two stars. This co-ordinate system is then rotated with the orbital motion of the two stars. In such a frame of reference the equipotential surfaces are almost spherical, close to the centres of the component stars. As the equipotential surfaces reach smaller values the surfaces become more distorted. Finally, for a certain value of the potential, the surfaces surrounding the two components touch at a point between the two stars. This is called the *inner Lagrangian point* and it is marked L_1 on the figure. The corresponding equipotential surface is called the *critical Roche surface*. It is composed of two *Roche lobes* surrounding the two components and touching at L_1. For smaller values of the potential, the two stars are enveloped by a common highly distorted equipotential surface.

The evolution of a massive binary star, to the point where it becomes an intense X-ray source, has been described by Van den Heuvel. The starting-point in the model is two stars undergoing the normal evolutionary process of hydrogen burning. One of the stars is about twenty times the mass of the Sun (20 M$_\odot$) and the other about 6 M$_\odot$. This starting-point is represented in Fig. 7.3(a) where the centre of mass of the system is also shown as a dotted line.

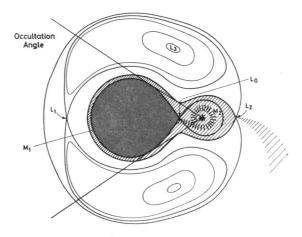

Fig. 7.2 The schematic diagram of equipotential surfaces of a binary system formed by a normal star M_1 and a compact collapsed companion M_2. L_o and L_1 are the principle (Lagrangian) points in the system where there is no net force in the rotating system. X-rays are emitted by the matter from the main star being accreted into the very strong gravitational field of the collapsed star

Naturally, the more massive star burns up its internal energy faster than the less massive companion. Eventually, a core of helium is produced by the fusion of between 15 and 40 per cent of the mass of the star. At this stage (see Fig. 7.3(b)) the helium core begins a rapid contraction and heating, which is accompanied by an expansion of the outer envelope which is still rich in hydrogen. In an isolated star a red giant would result from this expansion, however in the binary the expansion is limited as soon as the star fills its Roche lobe. At this point the outer layers of the star begin to spill over through the inner Lagrangian point L_1. Calculations by Paczyński and others show that this loss of mass tends to accelerate the rate of expansion of the envelope so that mass continues to transfer until almost the entire hydrogen-rich envelope is lost to the companion. A helium core is left (see Fig. 7.3(c)), which then begins helium fusion. Its radius subsequently shrinks to a few times the radius of

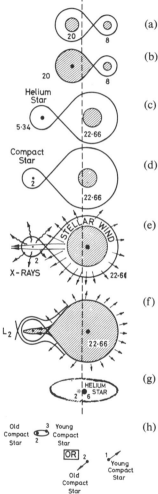

(a) $t = 0$, $P = 4^d.70$

(b) $t = 6.17 \times 10^6$ yr., $P = 4^d.70$ onset of first stage of mass exchange.

(c) $t = 6.20 \times 10^6$ yr., $P = 10^d.86$ end of first stage of mass exchange (beginning of first Wolf-Rayet Stage)

(d) $t = 6.76 \times 10^6$ yr., $P = 12^d.63$ helium star (= Wolf-Rayet star) has exploded as a supernova.

(e) $t = 10.41 \times 10^6$ yr., $P = 12^d.63$ the normal star becomes a supergiant; its strong stellar wind turns the compact star into a powerful X-ray source.

(f) $t = 10.45 \times 10^6$ yr., $P = 12^d.63$ onset of second stage of mass exchange; the X-ray source is extinguished and large mass loss from the system begins.

(g) $t \sim 10.47 \times 10^6$ yr., $P \sim 4$ hours onset of second Wolf-Rayet stage.

(h) $t \sim 11 \times 10^6$ yr., the helium star has exploded as a supernova; survival or disruption of the system depends on the mass of the remnant.

Fig. 7.3 Stages in the evolution of a massive close binary system, from the work of Van den Heuvel. The supernova explosion of the original primary star is assumed to have left a 2 M⊙ compact star. The dashed line indicates the centre of gravity of the system.

the Sun. Many examples of these helium stars are seen in binary systems. They are very hot and luminous and are probably very closely related to a class known as Wolf–Rayet stars.

The originally less massive star has now become the more massive component but it has not yet completed the hydrogen-burning phase of its evolution. The injection of the additional hydrogen-rich material rejuvenates the star, causing a prolonged lifetime.

The next stage in the evolution occurs in the helium star, which has a relatively small amount of nuclear fuel to supply its enormous luminosity. Computations of Arnett show that the successive stages of burning carbon, neon, oxygen and silicon never cause the radius to reach its Roche lobe. The star never loses any more mass and it continues to evolve as if it were an isolated helium star until the final collapse.

The final state reached in the evolution of a helium star depends very much on the 'initial' mass (after Roche lobe transfer). If the mass is more than 4 or 5 $M_\odot$, then a neutron star or black hole can be produced after a supernova explosion. In the case of a less massive helium star, it might well end up as a white dwarf. The evolutionary point in which a compact star has been formed is depicted in Fig. 7.3(d).

When the binary has evolved to a compact star with a hydrogen-burning companion, there is a relatively quiescent phase before substantial interaction between the two components once more activates the system. It might be expected that in the case of a neutron star being formed in the final collapse, this would be observable as a radio pulsar. However, the companion star provides a tenuous atmosphere, which is likely to absorb any radio emission, so that a binary pulsar of this type is probably undetectable.

It might also be expected that the production of an X-ray source would have to await the transfer of mass which occurs when the normal star fills its Roche lobe. It turns out that when this stage is reached the amount of material surrounding the X-ray source is probably so great that the X-ray emission is completely obscured. Most of the radiation would be degraded in energy and the large thick envelope would radiate mainly in the ultraviolet and visible regions of the spectrum to give the appearance of a normal star. The conditions for the binary to be an X-ray source are thus quite critical. The rate of transfer of material must not be too *great* nor too *small* otherwise the source will give insufficient emission to be detectable. Both requirements are met only during the brief stage between the end of hydrogen burning in the core and the time when the envelope has expanded to fill its Roche lobe. This period, of between 20,000 and 50,000 years, is the time when the star—having a mass of between 15–20 $M_\odot$—is a blue super-giant. The radius is between twenty and thirty times the radius of the Sun. Such stars are observed to be losing mass from the outer parts of their atmosphere at a velocity of about 1,000 km/sec. This stellar wind is driven by the radiation pressure exerted on the outer layers of very luminous stars. The rate of mass loss is about $10^{-6}\,M_\odot$ per year.

The compact star ploughs its way through the stellar wind and captures a small proportion of between 0.01 and 0.1 per cent of the material. Only a small proportion passing close to the compact star can be captured because of the high velocity of the wind. The bulk of the wind extends out into space without being affected by the companion and is lost from the system. The resulting accretion, about $10^{-9}\,M_\odot$ per year, is sufficient to power the X-ray source to a luminosity of 10,000 times the luminosity of the Sun in all wavebands. At the same time the stellar wind is of low density so that it is quite transparent to the X-ray emission.

We have now reached stage (e) of Fig. 7.3, which shows the evolution of a massive binary star. The compact star has now become a powerful X-ray source. This stage is, however, relatively short-lived in the total life of the system. Within 100,000 years the super-giant has filled its Roche lobe and the second stage of mass transfer begins. The X-ray source becomes hidden by the opaque cloud of material surrounding it. There is also substantial loss of mass from the system through the outer Lagrangian point L_2 as shown in Fig. 7.3(f).

Eventually the super-giant exhausts its supply of hydrogen and contracts to the helium-burning phase (see Fig. 7.3 (g)). The evolution of the star continues without further mass exchange until it, too, explodes in a supernova. The final state of the binary thus consists of two compact stars. This stage (see Fig. 7.3 (h)) may result in both compact objects remaining bound in a binary or the supernova explosion may disrupt the system, giving isolated stars which are separating with a high velocity.

The evolution of a massive binary in the above model predicts that all such stars will eventually go through a stage when they become strong X-ray sources. There are estimated to be about 6,000 or so such binaries in the Galaxy. The number of X-ray sources associated with massive binary stars is within reasonable agreement with expectation. The time spent in the X-ray-emitting phase is so short that, on average, only one in every 200 to 500 of the systems will be at this stage. This assumes that the formation of massive binaries is a continuous process in the Galaxy.

It must be remembered that the evolutionary model presented above may need considerable modification. A great deal of observational work is needed in the visible and ultraviolet regions of the spectrum so that the details of the mass loss can be determined. X-ray observations have already shown that compact objects exist in association with super-giant stars. Detailed study of the spectra and variability of the X-ray emission is needed in order to clarify the nature of the compact star, and the conditions in its immediate vicinity where the X-rays are produced and the spectrum modified by absorption. We shall describe in the following chapters what has already been established by the X-ray observatories and measurements in other wavebands. The next chapter describes the X-ray sources associated with either white dwarf stars or neutron stars and Chapter 9 describes the progress that has been made in establishing the existence of black holes associated with X-ray sources.

8 · White dwarfs and neutron stars

8.1 The death of stars

Throughout its life, from birth in a condensation of a gas and dust cloud, to its ultimate fate, a star evolves under the influence of gravitational forces, which are usually balanced by the pressures produced by the other forces of nature. In the last century the studies of the equilibrium conditions in stars had already shown that internal pressure must provide a means to balance the gravitational force which always acts so as to collapse the star. For a considerable period it was believed that the gravitational compression of stellar material, and the heat produced, was itself the source of energy which gives stars their high temperature and luminosity. However, the time for a gas cloud to collapse is now known to be much shorter than a star's lifetime. It was Eddington who, with remarkable insight, described the internal constitution of stars and suggested a 'subatomic energy source'. Shortly afterwards the details of the nuclear reactions, responsible for the prolonged life of a star, became clear. We have seen, in Chapter 6, how it is now possible to account for the main part of the evolution of stars by successive stages of nuclear burning. However, we must not give the impression that the theory of stellar evolution is understood in all its details. For example, as recently as 1979, Bok concluded an invited lecture with the remark that 'anybody who believes he completely understands even the formation of stars is a bloody fool'. Our limited knowledge about the birth of a star nevertheless does not prevent us from studying its ultimate fate. The details of the main sequence of evolution will no doubt also continue to stimulate astronomers for some time to come. The central problem for stars at the end of the evolutionary process has been recognized for many years. How can any system sustain itself against collapse when the mass and associated gravitational forces are increased more and more? In this and the following chapter we describe the observations of stars in their final states—after the internal nuclear fuel has been exhausted.

Three possible final states are now believed to occur—white dwarfs, neutron stars and black holes. In this chapter we describe the white-dwarf and neutron stars and in the next chapter we describe the progress made in proving the existence of the black holes, where gravity eventually overcomes the other forces of nature —in other words where the ultimate collapse finally occurs.

8.2 White-dwarf stars

The white-dwarf story began in 1844 when F. W. Bessel discovered that the motion of the 'Dog Star' Sirius could be accounted for by the presence of an unseen companion. The companion in this binary system was discovered by A. Clark in 1862, and is known as Sirius B. It is eleven magnitudes fainter than its bright companion Sirius A, but both components have the same surface temperature. The large difference in brightness must therefore be due to a difference in size. These white dwarfs were recognized by F. Bottlinger to be a new class of star. The unusual characteristics of Sirius B can be readily recognized by comparison of its properties with those of the Sun. Table 8.1 lists the parameters of the two stars.

Table 8.1

Quantity	Sirius B	Sun
Mass	$1.05 \, M_\odot$	$M_\odot = 1.989 \times 10^{33}$ g
Radius	$0.008 \, R_\odot$	$R_\odot = 6.96 \times 10^{10}$ cm
Luminosity	$0.03 \, L_\odot$	$L_\odot = 3.90 \times 10^{33}$ erg/sec
Effective temperature	27,000 K	5,800 K
Gravitational red shift	89 ± 16 km/sec	0.6 km/sec
Mean density	2.8×10^6 g/cm^3	1.41 g/cm^3
Central density	3.3×10^7 g/cm^3	1.6×10^2 g/cm^3
Central temperature	2.2×10^7 K	1.6×10^7 K

It is of course not possible to measure the central density and temperature directly, these values are inferred from particular models of the internal constitution of a star. There is no doubt, however, that these white-dwarf stars have average densities

1,000,000 times greater than the materials we normally encounter on the surface of the Earth. The measurement of the gravitational red shift predicted from Einstein's general theory of relativity confirmed that Sirius B has such a high density and small size. However, the existence of these compact stars posed a major problem for astrophysics, which was eventually solved by the application of the quantum-statistical theory of the electron gas developed by Enrico Fermi and P. A. M. Dirac in the mid-1920s.

The Fermi–Dirac theory showed that even the coldest electrons cannot all accumulate in the state of lowest energy—a consequence of the exclusion principle stated by Pauli. The total energy of an electron gas, and the associated pressure, must therefore be greater than zero even at a temperature of zero. It was realized almost immediately that this could explain the puzzle of the white dwarf. R. H. Fowler showed that the pressure of an electron gas, in which the low-lying states are successively filled, could be sufficient to support an object of about the mass of a star against its own gravitation. The radius of such a star turns out to be close to the measured radii of white dwarfs. The electron gas in such a star is in a state described as degenerate and for this reason white dwarfs are sometimes known as 'degenerate dwarfs'.

The theory of the degenerate electron gas was extended in the 1930s by Subrahmanyan Chandrasekhar who showed that if the effects of special relativity are included, then with increasing density the pressure provided by the electron gas is eventually insufficient to balance the force of gravitation. The mass of a white dwarf cannot therefore increase without limit. Chandrasekhar found that the critical mass, above which no stable white dwarfs could exist, is $1.44 \, M_\odot$. The establishment of the 'Chandrasekhar limit' was a major discovery and immediately presented a problem for the final state of stars that are unable to shed enough matter, in their evolution, to remain below the limiting mass for a white dwarf. As we have seen, the supernova explosion associated with the collapse is the inevitable result. Later, in this and the following chapter, we will discuss the collapsed objects that remain after these dramatic events.

The conditions inside white dwarfs began to be understood in the 1940s. The first step came with the recognition that the theories on the electron thermal conduction of metals could be applied to the interior of white-dwarf stars. It was found that the thermal conductivity was so high that the core of a white dwarf is at an almost uniform temperature. However, the surface layers are in a quite different, non-degenerate, state. This thin surface layer provides a barrier which prevents rapid transfer of the internal heat of the star. In 1952 Leon Mastel pointed out that the thermal energy contained in the interior of the white dwarf, and leaking slowly through the thin outer insulating layers, provides sufficient energy to account for the luminosities observed for white-dwarf stars.

By 1961 the theory of the internal structure had been refined by Edwin Salpeter to include the interactions of the nuclei with the degenerate electron gas. These corrections, such as those arising from attractive forces, serve to reduce the energy and the pressure of the interior material. This, in turn, is reflected as a reduction in the maximum allowable mass of white dwarfs to about the same mass as the Sun. The exact value depends on the temperature of the material. Detailed theoretical work on the internal constitution of white dwarfs still continues; however, this is beyond the scope of our discussion.

8.3   Neutron stars

The idea of a neutron star dates from the discovery of the neutron itself. When word came from Cambridge of Chadwick's identification of the new particle, Bohr and Landau, then in Sweden, discussed the implications of this great discovery. Landau first suggested the possibility of cold dense stars composed primarily of neutrons. Baade and Zwicky independently proposed the idea of neutron stars which would be of extremely high density and small radius. Zwicky also suggested that they might be produced in a supernova explosion, as an end-point in the evolution of massive stars, the gravitational collapse providing for the release of tremendous amounts of energy.

Oppenheimer and Volkoff made the first calculation of the properties of neutron stars in 1939. They assumed the matter to be composed of a dense gas of free neutrons. The radiation resulting from such stars as a result of their temperature appeared too weak to be observable at typical stellar distances, because of their small size. Further theoretical work proceeded slowly until the discovery of the radio pulsars in 1967 by Bell and Hewish.

Although there are now known to be more than 300 radio pulsars in the Galaxy, progress on understanding their nature has been slow because only in one case does the pulsar exist in a binary system. This binary pulsar, discovered by Hulse and Taylor in 1975, has been studied extensively and has begun to reveal much valuable data on the masses of the two components, the effects of general relativity on the orbital motion, and the beaming of the radio pulsations.

Gold first suggested that the radio pulsars are rotating neutron stars, which have the axes of their magnetic field aligned at an angle to the axis of rotation so that beamed radio radiation sweeps out, giving the characteristic radio pulses. Subsequently, theoretical work on the properties of neutron stars has been intense. The discovery of compact X-ray sources in binary systems has provided further stimulus, with the additional possibility of testing the predicted characteristics derived from the theoretical models of neutron stars.

The early calculations on the internal structure of a neutron star, made by Oppenheimer and Volkoff in 1939, assumed the star is composed of a sea of non-interacting neutrons. The neutrons behave like the degenerate electron gas of a white dwarf but the pressure provided can withstand much greater gravitational forces. Densities of above 10^{14} g/cm^3 can be achieved compared with the limiting density of between 10^7 g/cm^3 and 10^8 g/cm^3 for white-dwarf material.

More recent calculations indicate that the internal structure and equation of state are likely to be more complex than the simple model of Oppenheimer and Volkoff. Fig. 8.1 illustrates the internal structure derived from a model produced by Pandharipande and co-workers in 1976. The composition of the neutron star in the outermost regions is expected to be the same as the interior of a white-dwarf star: nuclei of iron surrounded by a sea of degenerate electrons. The density at the surface is about 10^4 g/cm^3. The iron nuclei are expected to form a solid crystalline crust which is very rigid—much more rigid than ordinary solid materials because of the higher energy of interactions between the nuclei.

Beneath the solid crust, the density rises very rapidly and reaches a maximum at the centre of the star. The models are computed from an equation of hydrostatic balance, which must also include the effects of general relativity in the intense gravitational fields encountered in the interior of neutron stars. When account is taken of the general relativistic effects, it is found that the pressure increases at a greater rate than would be the case when moving to the interior of a non-relativistic assembly of material. In order to solve the equation it is necessary to know the relation between the pressure and the density of the material. We are familiar with such equations of state for normal materials; for example the laws for perfect gases

(Charles and Boyle) are an essential requirement for a first course in physics. At densities of normal materials the equations of state are fairly well known but there is considerable uncertainty when the material approaches the density of nuclear material. The equation of state determines the maximum mass of the neutron star. An increase in mass increases the central pressure and at some point the material will be unable to provide this pressure. An absolute upper limit to the mass of a neutron star has been determined by Rhoades and Ruffini in 1974. Their value of 3.2 $M_\odot$ assumes the most extreme forms for the equation of state. Ruffini also pointed out that it is hopeless to try to establish the effective value of the critical mass, or its other parameters, by direct theoretical arguments. Progress can be made only by collection of experimental data on neutron stars and by direct comparison with existing theoretical predictions. The discovery of neutron stars as the sources of intense X-radiation has helped to begin the testing of these theoretical predictions.

8.4 X-ray emission from the compact objects

We have remarked in earlier chapters that the discovery of the faint optical counterpart of the strongest continuous X-ray source, Sco X-1, immediately suggested that an unusual object must be responsible for the remarkable properties of this source. The suggestion

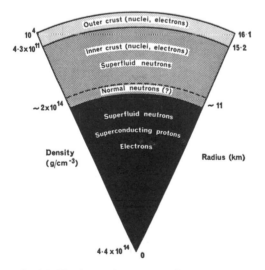

Fig. 8.1 The internal structure of a neutron star

that the accretion of material on to the surface of a compact star was the source of the high X-ray luminosity can be understood in terms of conversion of the kinetic energy of the in-falling material into heat. If the gravitational potential at the surface of the star is θ then the luminosity produced will be given by:

$L \sim \theta \dot{M}$ where $\dot{M}$ is the rate at which material falls on the surface.

In the Newtonian approximation the gravitational potential becomes GM and the luminosity relative to the Sun is then:

$$L\big/L_{\odot} \sim 5 \times 10^7 \frac{(M/M_{\odot})}{(R/R_{\odot})} \dot{M}$$

and the temperature:

$$T \sim 10^7 \frac{(M/M_{\odot})}{(R/R_{\odot})} \text{ K}$$

where M is in units of solar masses per year ($M_{\odot}$/year). For typical white dwarfs $(M/M_{\odot})/(R/R_{\odot}) \sim 10$, and one obtains a temperature of 10^8 K and $L/L_{\odot} \sim 100$ when $M \sim 2 \times 10^{-7}$ $M_{\odot}$/year. These values can reasonably explain the luminosity and temperature of the X-ray emission from Sco X-1. The idea that Sco X-1 consists of a white-dwarf star as a companion to an ordinary star, with mass flowing from the latter, appeared quite plausible.

In spite of more refined calculations of the accretion phenomenon, the nature of the compact object in the Sco X-I system is still not clear. That the source is in a binary system has been established through careful measurements of the visible counterpart. No periodic signal has yet appeared in the X-ray data and progress has been confined to measuring the X-ray spectrum and comparing its variations with the random fluctuations in the visible signals from the system. Eventually, these studies may be able to distinguish between a white dwarf and a neutron star as the compact object which provides the extreme conditions necessary to give the copious X-ray emission.

Fortunately, a considerable number of the X-ray sources in the Galaxy provide much greater information about the mechanism of their X-ray production and these data can unambiguously discriminate between the possible different types of compact star.

The discovery of the pulsating (4.8 sec) and eclipsing periods (2.1 days) of the Cen X-3 source marks the start of a very fruitful line of research. Information on the duration of the X-ray eclipse, the Doppler shift of the pulsations, and long-term changes in the intrinsic pulse period can lead to valuable information on the nature of the compact object and its companion star.

By 1978 a large number of the galactic sources had been studied in order to detect regular variability. The shortest period of 33 msec is associated with the neutron-star rotation period and the rotational energy supplies the source of the high-energy particles, which provide the characteristic radiation of the nebular itself. The longest period observed with certainty is the thirty-five-day cycle associated with the Her X-1 source. Fig. 8.2 illustrates the observed regular periods in the galactic sources, and also indicates the likely mechanism for providing the clock. The means by which the X-radiation is generated is also shown.

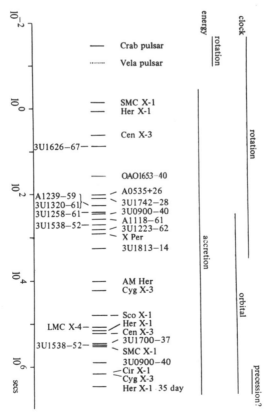

Fig. 8.2 Regular periods observed in X-ray sources

The discovery of such a wide range of variability has had an important effect on other branches of astronomy. Unlike the radio pulsars, the techniques of traditional astronomy are of vital importance in the investigation of the X-ray sources. The eclipsing source Vela X-1 (3U 0900–40) provides the best example of the interaction of X-ray investigations and optical studies.

The companion star to Vela X-1 is a bright star (HD 77581) and it is therefore relatively easy to measure the Doppler motion of its lines in the visible region of the spectrum. There are some fluctuations in the measured values, which are probably caused by variations in the strong stellar wind emitted by the star. In 1976 workers from MIT discovered that the associated X-ray source, in addition to its known regular eclipses of 8.9-day period, also exhibits regular pulsations with a period of 283 sec. For the first time it became possible to apply Kepler's laws to the system and obtain the ratio of the masses of the two stars. The duration of the X-ray eclipse also enabled the size of the primary star to be estimated. The analysis of the data gave a mass for the compact object that lies in the range 1.2–2.4 $M_\odot$. For the first time it was conclusively proved that the compact object must be a neutron star because the mass was greater than the allowed value of a white dwarf.

Three other X-ray sources have regular eclipses and pulsations; the source in the Small Magellanic Cloud SMC X–1, Cen X–3 and

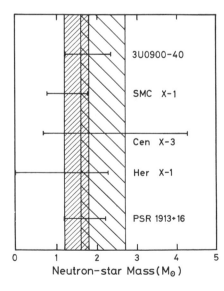

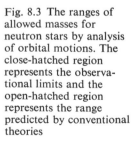

Fig. 8.3 The ranges of allowed masses for neutron stars by analysis of orbital motions. The close-hatched region represents the observational limits and the open-hatched region represents the range predicted by conventional theories

Her X-1. The analysis of these systems is made more difficult by the faintness of the optical counterparts; however, limits in the masses of the compact objects in these systems are useful and these have been plotted on Fig. 8.3. The estimates for the mass of the neutron star in the binary radio pulsar have been included. In 1979 Taylor and his colleagues concluded that this system probably consists of two neutron stars in orbit, each having a mass of about 1.4 $M_\odot$. Formally, the limits in the mass of the radio pulsar range from 1.3 to 2.3 $M_\odot$. If we make the assumption that the neutron stars all have the same mass, then this is very close to the limiting mass given by the conventional theories of many-particle physics.

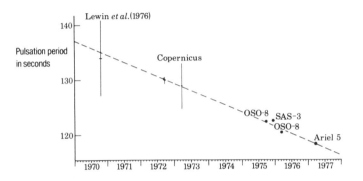

Fig. 8.4 The changing period observed by various workers for the X-ray source GX 2 + 5. The increased rotation is presumably caused by transfer of momentum from the in-falling gas stream

Additional insight on the nature of the compact object is gained by observing how the period of pulsation changes over an interval of many years. The most extensive measurements are available on the source designated GX 2+5. Measurements are available from 1970 and show that the pulsation period has decreased by more than 10 sec, from its value of just over 130 sec, when the first measurements were made. The measurements of the pulse period are shown in Fig. 8.4. There has been no report of the detection of orbital motion in this system, however this may indicate that the orbit of the system is inclined to the line of sight, so that the orbital motion would be undetectable.

The significant change in period for the source, which is about 10 per cent over an interval of seven years, is enormous when compared with the changes in periods that have been measured for the radio pulsars ($\sim$ one part in 10^7 per year). In the case of the radio pulsars the lengthening period is the result of a decrease in the rotational energy, which gives rise to the radio pulsations themselves. However in the case of the X-ray pulsators, the period is *decreasing* and the compact object is therefore *receiving* energy, as is evident from the increase in the rate of rotation. The most likely mechanism responsible for the spin-up of these X-ray sources is the conversion of momentum of a gas stream into angular momentum of the neutron star. The windmill is a familiar example of the conversion of linear to angular momentum.

It turns out that the changes are so great that it is impossible to transfer sufficient angular momentum, in the case of a white dwarf, to bring about such a large rate of change in period. However, the much smaller neutron star can have its rotational period changed by these large amounts. The measured values of fractional changes in period are shown in Fig. 8.5. Theoretical predictions on the changes in period expected for a given luminosity (which is related

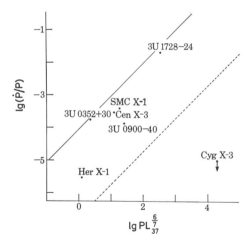

Fig. 8.5 The relative rate of change of the period of X-ray pulsars plotted against their luminosity $\times$ period. The solid line represents the theoretical limiting values for neutron stars and the dashed line the limiting values for rotating white-dwarf stars

to the accretion rate and therefore the maximum torque available) are also shown on the graph. The solid line gives the limit above which the changes could not be induced by the accretion process. The broken line gives a similar limit for the white-dwarf stars. The six sources for which measurements are available all lie above the maximum value possible for white-dwarf stars, which implies that the compact objects in these X-ray systems are neutron stars.

Further evidence for the existence of neutron stars in binary X-ray sources was obtained in 1977 by workers led by Joachim Trümper of the Max Planck Institute at Munich. It is well known that when a star collapses in a supernova the conservation of the magnetic flux will lead to extremely high magnetic fields at the surface of a neutron star. Fields of between 10^{10}–10^{12} Gauss are predicted. For comparison the field at the surface of the Earth is about 0.33 Gauss. It is this extremely high magnetic field which, when misaligned to the rotational axis of a rapidly rotating neutron star; is believed to provide the radio pulses observed from the pulsars.

The workers from the Max Planck Institute studied the Her X-1 source with instruments carried on high-altitude balloons. Their discovery of an X-ray spectrum line at 53 keV can be interpreted as evidence for the very high magnetic field in the region, giving rise to the hard X-ray emission. The fact that the strength of the line varies in phase with the X-ray pulsations at lower energies is convincing proof that the origin must be strongly linked to the magnetic properties of the X-ray source. The narrowness of the line can be used to estimate the size of the region of emission. Trumper and his colleagues estimated that the region could be as small as a few square centimetres in cross-section and extend to a height of as much as one kilometre above the surface of the neutron star. Their estimate of the magnetic field (5×10^{12} Gauss) is in good agreement with the theoretical predictions and inferred values of the magnetic fields of radio pulsars.

The prospect of probing the surface conditions of neutron stars by such measurements is an exciting one and there will be great efforts made to refine the instruments so that the detailed shape of the line, and its energy, can be determined more accurately. The discovery has also been a great stimulus to theoretical work being done on the conditions necessary to give this surprising phenomenon.

Our knowledge of the galactic X-ray binary sources, and the nature of their compact companions is, as we have seen, definitive in only a handful of the two hundred or so strong point-sources of X-ray emission in the Galaxy. In the last chapter we described the evolutionary path of a binary system, where in the X-ray-emitting phase the optical counterpart is a super-giant star. There are, however, several examples where the optical counterpart is a lower mass star. The notable case is the Her X-1 source. In this system the low-mass visible star, being less luminous than the super-giant stars, has its optical characteristics considerably perturbed by the intense X-radiation emitted from its companion neutron star. When the

Table 8.2 Characteristics of some Stellar X-ray sources

Source	Mag	P	Pulsing?	Mass$_{op}$	Mass$_x$	Spectrum	Remarks
Early super-giant binaries							
Cyg X-1 = HD226868	9	5.6^d		<25:	<14	09.7 Iab	Primary mass assumed
Vela X-1 = HD77581	6*	8.9	283^s	22	1.6	Bo.5 Ib	
Cen X-3 = Krzeminski's star	13*	2.1	4.8	17	0.7	BO Ib-II 06.5III	
SMC X-1 = Sk 160	13*	3.9	0.7	15	1.1	BO Ia	
3U 1700-37 = HD153919	6*	3.4		27	1.3	O6	
4U 0900-40 = HD77581	7	9.0^d	283^s	24	1.6	BO.5 Iab	
4U 1223-62 = WRA977	10	$22.^d6$, $40.^d8$?	697^s	30	?	B2 Ia	
Low-mass binaries							
Her X-1 = HZ Her	15*	1.7	1.2	2.2	1.3	A-F var	
Cyg X-2 = G sub dwarf	15	0.9?		1.5?	1?	A-f var	
Sco X-1 = V818 Sco	13	8–12?		1.5?	2–3		
		0.79		1	1.3:	H,He II em	
3U 1809 + 50 = oM Her†	13 var	0.13		0.4:	1:	H,He II, He I em	
Cyg X-3 = ?	*	0.2		1?	1?	Not seen optically	
Be stars							
3U 0352 + 30 = X Per†	6	560?	835	20:	40:	BOe	
		0.9	835	20:	1:		
MX0053 + 60 = γCas†	3			25	<3	BOe	No evidence this is a binary; masses are upper limits.
A0535 + 26 = HD245770	9	>17	104			Be	X-ray pulsing gives $P>17^d$; no known optical P; X-ray nova
Transient X-ray source							
A0535 + 26 = HD245770	9	>17	104			Be	See above; no optical outburst
A0620-00 = Nova Mon 1975	12–20	4?				H em	Optical outburst with X-ray event
3U 1118-61 = ?	12	?	405			BOe	Identification not certain

* Eclipsing in X-rays.
† Low X-ray luminosity.

X-ray source is eclipsed by the star, we observe the characteristics of a normal cool (F-type) star; however, when the X-ray-heated side of the star comes into view the atmosphere is modified to such an extent that it takes on the characteristics of a much hotter A star. In Table 8.2 we have listed the properties of some of the stellar X-ray sources which have been identified with optical counterparts.

Of the four classes of binaries, those containing an early super-giant have been studied most successfully. The difficulties encountered in observing the low-mass binaries can be recognized by the faintness of their optical counterparts, which range from thirteenth to fifteenth magnitude and in the case of Cyg X-3 the counterpart can be seen only in the infrared part of the spectrum. The Be stars are also difficult to study as they generally rotate at high rates, which makes it difficult to obtain the details of their orbital motion because their spectrum lines are broadened. The transient sources, by their very nature, require very rapid response on the part of observers and the X-ray and optical observatories cannot always be made available at short notice.

The galactic X-ray sources will almost certainly continue to provide a rewarding subject for investigation for many years. Speculation about the nature of the compact objects in many of the binaries is continuing and we may eventually find that white dwarfs can also provide the conditions necessary to account for the observed X-ray emission from some objects.

9 · Black holes

9.1 Singularities

We are now well adjusted to the concept that the Universe began with a big bang. In the following chapters we will probe some of the observations and outstanding problems associated with the evolution from this initial singularity. The controversy that raged before acceptance of the big bang theory is a measure of the distaste felt by many scientists for a singularity, in this case creation. What preceded it presents a problem which can, perhaps, only satisfactorily be dealt with by theologians and philosophers. It has also been hard for many scientists to digest the peculiarities associated with the singularity of a black hole. Science-fiction writers have nevertheless embraced the black hole with enthusiasm and even entered the singularity with the words of Captain Kirk: 'to boldly go where no man has gone before'. Unfortunately, most astronomers are unable to split this particular infinitive so easily. We must concentrate instead on the theoretical explanations for the phenomenon of the black hole and the observational evidence for their existence.

9.2 The critical mass for gravitational collapse

The first detailed treatment of the equilibrium conditions for neutron stars was made by Oppenheimer and Volkoff in 1939. As we have seen in the last chapter, the major conclusion reached in their classic paper was that stable neutron stars can exist only if the mass lies in the range between 0.1 and 0.7 $M_{\odot}$. The assumptions made were also very clear. The internal structure of the neutron star is described by an equation obtained from the properties of a gas of neutrons up to relativistic energies and the large-scale structure is described on the basis of the equations of general relativity, applied to matter distributed as a perfect fluid.

It is remarkable that the major conclusions of Oppenheimer and Volkoff's work have remained valid, even after thirty years of refinement to the calculations, and much critical review of the initial assumptions. Apart from a change in the value of the critical mass (to 3.2 $M_\odot$) it has remained inevitable that gravitational collapse will occur once this critical mass is exceeded. Fig. 9.1 shows the ranges of the equilibrium conditions for the white-dwarf and neutron stars, and the regime of density and mass for gravitational collapse.

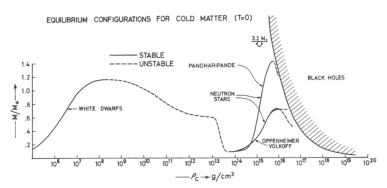

Fig. 9.1 The possible equilibrium conditions for collapsed matter. The masses (in terms of the solar mass $M_\odot$) are plotted as a function of the central density of the collapsed object. The solid line represents regions of stable equilibrium for degenerate matter. Limiting masses for white dwarfs (the Chandresekhar limit) and neutron stars are shown. The value in the limiting mass for neutron stars will influence the ratio of neutron stars to black holes to be found in nature

It was also Oppenheimer who, together with Snyder, first discussed the properties of a star undergoing gravitational collapse. Their paper, published in 1939, described how the energy emitted outward from the surface of the star will be reduced by three effects:

The Doppler effect from the receding source, the large gravitational redshift and the gravitational deflection of light which will prevent the escape of radiation except through a cone about the outward normal a progressively shrinking aperture as the star contracts. The star thus tends to close itself off from any communication with a distant observer; only its gravitational field exists.

A great deal of research on the properties of these collapsed stars has been carried out since that time but once again the main conclusions of Oppenheimer have remained valid.

In order to deduce the properties of a collapsed star, we are forced to use the general relativistic theory of gravitation. The most important property of a black hole can, however, be understood in terms of the familiar Newtonian approximation. As long ago as 1798 Laplace wrote:

> A luminous star, of the same density as the Earth, and whose diameter would be two hundred and fifty times larger [in mass] than the Sun, would not, in consequence of its attraction, allow any of its rays to arrive at us; it is therefore possible that the largest luminous bodies in the universe may, through this cause, be invisible.

Although not an exact analogue, the application of general relativity leads to a similar result. For a star such as the Sun a distance known as the Schwarzschild radius defines a limiting sphere. If all the mass were to be confined to within this sphere, its radius would define the 'event horizon', a critical surface from which no radiation can escape. For the Sun this radius is 2.95 km—much less than the actual solar radius. For the Earth the critical radius is only 8.86 mm. This characteristic distance, and the region outside, is described by a geometrical form known as the Schwarzschild metric. (The 'metric' describes the local curvature of space-time in modern theories of gravity.) Inside, the region is described by a different metric. It might be thought that the event horizon is as unphysical as the apparent singularity in the Newtonian formula for the gravitational potential ($-GM/R$). However, this only applies to the region outside a massive body, and so avoids the infinity at the point where the radius becomes zero. Calculations show that a star with greater than the maximum allowable mass for a neutron star can eventually shrink inside its Schwarzschild radius. Collapsed matter may also exist even though it does not exceed the limiting mass of a neutron star. Such mini black holes may have been created during the initial stages of the formation of the Universe.

9.3   The properties of black holes

Three theorems about black holes have been proved under fairly general conditions. The first is that gravitational collapse always ends in a singularity of some sort, even if the collapsing body is not spherical before the collapse. The second theorem is that the rest of the Universe is always shielded from the singularities by the development of an event horizon, which prevents matter and radiation from escaping. The third theorem is that once a black hole has been formed, the only properties which are observable outside the event horizon are the mass M, the angular momentum J, and the electric charge Q. All the other properties of the body—its shape, mechanical strength and magnetic multipole moment are destroyed in the collapse. The ability to characterize a black hole fully by these three parameters M, J and Q led to J. A Wheeler's quotation: 'Black holes have no hair.'

Recently, the Cambridge astrophysicist Stephen Hawking made the discovery that black holes are, however, not completely black. Even in a vacuum, particles of matter— and corresponding particles of anti-matter—are being continuously created; these opposite particles annihilate each other almost immediately, but the process is unobservable since the energy of creation is exactly equal to the energy of destruction. However, in the vicinity of the event horizon of a black hole, it is possible in the intense gravitational field for one of these particles to enter the hole—leaving the other to escape with a lot of energy. If this energy escapes from the region outside the black hole, as a gamma ray for example, then it can be obtained only by a small decrease in the mass of the black hole. For a black hole of about the same mass as the Sun, this process would eventually lead to its evaporation in a time of 10^{67} years. Such black holes can therefore be considered to be stable within the lifetime of the Universe. For a small black hole the evaporation proceeds more rapidly, so that if there were black holes of about 10^{11} kg formed at the initial big bang, then by now these should be completing their evaporation with a flash of gamma rays. Searches are already underway to find the numbers of these mini black holes, if any, in our Galaxy.

The slow evaporation of black holes, formed at the end-point of the evolution of a star, would produce so little radiation that they would be quite unobservable. How then are we to detect a stable black hole with confidence, if no radiation or particles can ever reach us because of the event horizon? Fortunately a black hole has one property—its gravitational field—which gives us the chance of proving its existence and investigating the region near its event horizon. An obvious way of detecting a black hole would be to observe the bending of light from a star as it passes through the intense gravitational field. This method was the first successful test of the theory of general relativity where the light from a star was deflected by the predicted amount in the gravitational field of the Sun. Unfortunately, it is very improbable that a black hole would line up in front of a star to deflect its radiation, although in this unlikely event the black hole would act as a gravitational lens to provide an unusually magnified object.

The most direct method for the detection of black holes is by observing their gravitational effects on other bodies. For this reason the binary stars that have unseen, yet massive, companions provide the most likely conditions. Several searches of the star catalogues have been made in the attempt to select candidate systems for more intensive study. Unfortunately, the presence of a massive unseen companion is not sufficient evidence for a black hole in a binary system. If the visible star is particularly luminous then it may swamp the emission of any optical light from its companion, which, although undetected and beyond the mass of a neutron star, is in fact not a compact object.

The discovery of X-ray sources in binary stars have provided what might well be the most powerful means of detecting black holes. If the companion to a normal star is a compact object, then we have seen that large amounts of energy can be converted into radiation when the mass flowing from the normal star falls on to the surface of the compact object. This is the basic process that provides the energy radiated from neutron stars and possibly white dwarfs, where the energy appears in the form of intense X-ray emission. However, for a black hole we encounter an immediate problem, as although the particles reach extremely high velocities as they approach the event horizon, they pass through it at the speed of light and from there on are invisible to us. How then is it possible to observe any radiation through the accretion of matter by a black hole?

The apparent paradox of copious radiation from a large black hole was, in fact, resolved in 1962 by the Russian theorist L. S. Shlovskii. He pointed out that the collisions between particles that are approaching the event horizon can release enormous amounts of energy—as much as 10 per cent of their rest-mass energy (0.1 mc^2). This energy is large compared with that available from nuclear fusion (0.007 mc^2). Even if all the particles were falling towards the centre of the compact object, which would not be the case in a rotating binary system, there would inevitably be some collisions as the material became compressed. It is more likely, however, that particles would arrive in the region of the compact object and become trapped in orbit, rather than falling directly towards its centre. Under these conditions, there would be a greater likelihood of collisions within the disc that forms around the object. The inner portions of the disc would be moving at greater rates than the outer portions, so considerable collisional heating would occur. It is the creation of this accretion disc and the radiation emitted from it that provides the X-ray signature of the black hole.

9.4 Cyg X-1: The first black-hole candidate

The X-ray source Cyg X-1 was discovered by the NRL group in 1966. A comparison of the measurements carried out over the period between 1966 and 1968 indicated that this source was the first to be obviously variable and consequently likely to be rather small compared, for example, with the dimensions of the supernova remnant X-ray source which had been identified as the Crab Nebula.

When the Uhuru satellite began making observations, it was natural that Cyg X-1 would be studied intensively because of its known variability. Although the variability was immediately confirmed by the Japanese astronomer M. Oda and the team of scientists working with the data, the search for regular pulsations characterizing the X-ray source as a neutron star was fruitless. Nevertheless, the discovery of fluctuations of the source on a timescale of 50 msec confirmed that the X-rays must be coming from a very small region. This result was important, for it promoted the search for pulsations from other sources leading to the many discoveries described in the previous chapter.

Cyg X-1 remained an enigma. There were no regular eclipses and no regular pulsations. Eventually, the position of the source was refined so that optical astronomers could begin to look for likely counterparts to the X-ray source. The bright star HDE 226868 was sufficiently close to the area of uncertainty of the X-ray source (as described by the Uhuru satellite), and optical astronomers began to study its properties. The area of uncertainty (from the second Uhuru catalogue) and the star field are shown in Plate 14. Astronomers from the Royal Greenwich Observatory, Louise Webster and Paul Murdin, discovered that this candidate star was taking part in binary motion, a result confirmed almost immediately by C. T. Bolton. The measurements of the Doppler shifts of the visible spectral lines, indicating this binary motion, are shown in Fig. 9.2. The analysis of the velocity curve of the optical candidate proved to be very exciting.

Although the mass of the unseen companion could not be estimated accurately, there were strong indications it must have a mass greater than that of a neutron star.

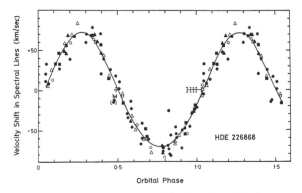

Fig. 9.2 Measurements of the Doppler shift in the optical spectrum lines in the primary star associated with Cygnus X–1

The evidence that would associate the optical candidate with the X-ray source became of great importance. The first link came indirectly through the discovery of a radio source by Braes and Miley, which almost certainly had to be associated with the optical star, since their positions coincided to within one arc second. When data from the Uhuru satellite were compared with the radio signals, it became evident that there was an association of the radio star with the X-ray source. At the same time as the X-ray source reduced in strength, the radio source increased in strength from below the threshold of detection. These anti-correlated changes, and the coincidence in position of the radio source with the optical star, suggested that the X-ray source was also associated with the binary star.

Confirmation that the optical candidate was associated with the X-ray source came accidentally through observations made in 1972 with the X-ray telescopes on the Copernicus satellite. Shortly after the launch of the satellite it became clear that these telescopes could obtain the positions of X-ray sources with much greater precision ($\sim$ 10 arc seconds) than had been appreciated when the instrument had been designed. The stability of the control system of the satellite was so good that, even in the absence of a star to guide the telescopes, the pointing direction of the telescopes could be maintained to a few arc seconds.

The Cyg X-1 source was the first available source that could be used to calibrate the pointing-directions of the X-ray telescopes, relative to the optical sensors, after leaving a bright guide star. The method used for calibration consisted of making a spiral scan of the region around the expected position of the X-ray source. The scan of the Cyg X-1 is shown in Fig. 9.3. The results of this experiment were puzzling, because at the start of the spiral scan there was almost a complete absence of signal. Many subsequent tests ensured that there was no abnormality in the centre of the telescope response, that the satellite was pointing at the expected position, and it did not require an appreciable time to settle after moving from its guide star. An abnormality in the X-ray source itself was the only possible explanation.

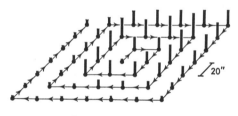

'Spiral' scan of Cyg X-1

Fig. 9.3 The data (0.5–2.0 keV) obtained from the Copernicus satellite as it scanned the region containing Cygnus X–1. The length of the bars indicate the strength of the signals at each pointing position. Note the absence of a strong signal at the centre of the scan pattern

With the assumption that Cyg X-1 and the optical candidate were in the same binary system, Paul Murdin calculated that the time of the observation was quite close to the time when the X-ray source would be most likely to be obscured by the primary star. Further observations were made with the Copernicus satellite which confirmed that the signals from the X-ray source were, at times, partially obscured when the X-ray source passed close to superior conjunction. The most dramatic of these events observed with the Copernicus satellite is shown in Fig. 9.4 where the X-ray signals are, at times, almost completely cut-off for intervals as short as a few seconds. These results confirmed that the X-ray source was associated with the binary star and that the size of the X-ray emission must be quite small. Other satellite measurements have confirmed these observations of absorption of the X-rays by a gas stream passing from the main star to the X-ray source.

With the positive identification of Cyg X-1, in the binary system containing the optical star HDE 226868, this system became the subject of intense study. It became essential to prove that the optical star was not unusual in any way since if, for example, it could be shown that its distance from the Earth was considerably less than assumed, this would reduce the value for its luminosity and therefore its mass. The unseen companion would then also have a reduced mass—below that of a black hole. Other suggestions were also put forward which would not require a black hole for the companion.

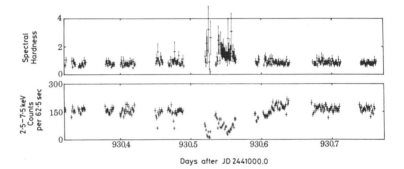

Days after JD 2441000.0

Fig. 9.4 The 'absorption dip' observed in the signal from Cygnus X–1. The spectral hardness is a measure of the amount of gaseous material providing the decreases in signal. Discovery of these absorption dips in phase with the optical star HDE 226868 confirmed the identification of the X-ray source with the binary star

Even a three-body system was suggested to provide sufficient mass in the system without invoking a black hole. The twisted-magnetic-field model was also suggested as a way of producing the intense X-ray emission. None of these suggestions is now taken seriously.

The orbital period of the optical star and its Doppler-shifted spectral lines gives a value for the mass-function (see Chapter 7):

$$\frac{M_x^3 \sin^3 i}{(M_x + M_s)} 2 = 0.23 \text{ M}_\odot$$

where M_x and M_s are the masses of Cyg X-1 and HDE 226868 respectively. For this system the mass-function has been determined entirely from the optical observations, since there are no periodic signals from the X-ray source. The estimates for the mass of the optical star, and the inclination of the system, are thus of importance if we are to obtain the mass of the X-ray source. The best estimate for the mass of the optical star is about 15 M $_\odot$. If we then solve the equation for the mass of the secondary as a function of the inclination of the system, we obtain the mass of the unseen secondary for two possible values for the mass of the primary. This is shown plotted in Fig. 9.5.

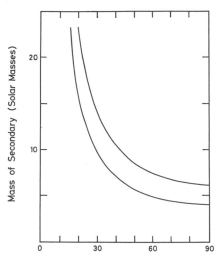

Fig. 9.5 The allowed mass of the secondary object as a function of orbital inclination. Two values of the primary star are assumed. The upper curve corresponds to 30 M $_\odot$ and the lower curve to 15 M $_\odot$. The secondary star exceeds the mass of a neutron star for all but the most extreme assumptions and is therefore likely to be a black hole

The suggestion that the optical star might be nearby, less luminous, and therefore less massive, had to be considered carefully. Fortunately, it was possible to remove any ambiguity about the distance and therefore the luminosity, by the traditional methods of optical astronomy. Bruce Margon and others were able to estimate the distance of the star by observing the amount of interstellar absorption of a large number of stars in the region containing the Cyg X-1 source. The relationship between the distance and the amount of interstellar reddening was established, and it was shown conclusively that the distance, and therefore the luminosity, of HDE 226868 is quite normal.

The suggestions of a three-body system or of X-ray emission resulting from annihilation of magnetic fields appears to be ruled out by the observation of the absorption events discovered with the Copernicus satellite. These events are phased with the known period of the optical star. The complete absorption of the X-ray emission by a gas stream would seem to remove the possibility that interaction of the magnetic field between two normal stars could provide the X-ray emission. It would be unlikely that the emission would be constrained to a very small region, as implied by the short duration of some of the absorption events.

The controversy about the nature of the X-ray source in the Cyg X-1 system continues. There is some uncertainty in the estimates of the mass of the primary star, even though its distance is fairly well determined. These mass estimates are based on the observations of the optical spectrum and these can be in error by as much as a factor of three, according to some workers. Even so, it appears that the mass of the primary must be greater than $10 \, M_\odot$, which gives a secondary having a mass greater than the value of $3 \, M_\odot$, which is beyond the expected limit for the mass of a neutron star.

Another line of evidence for the black hole in the Cygnus X-1 source is based on the observations of very rapid fluctuations in the X-ray signals from the source. The Goddard Space Flight Center Group have investigated these fluctuations with the aid of large proportional counters flown on rockets. There have also been more recent measurements with large counters installed on the first NASA High Energy Astrophysics Observatory. Rothschild and his colleagues at the Goddard Space Flight Center observed bursts of X-rays separated by intervals as small as 5 msec. These bursts are interpreted as arising from hot spots, orbiting very close to the event horizon of the black hole.

Observation of the pulses, which are not regular as in the case of neutron stars, is the most promising way of investigating the characteristics of the intense gravitational field close to the boundary of a black hole. It has, for example, been predicted that the fluctuations will be more rapid if the black hole is rotating. This line of attack is very similar to the indirect approach used in the study of elementary particles of nuclear physics (the properties of a particle are inferred by the results of scattering experiments). For the black hole we expect a clear cut-off in the time signature of the signal when it is equal to that of the time of revolution of the last stable orbit.

When we consider the available evidence in support of the black hole in the Cyg X-1 source, we are forced to conclude that the efforts of the devil's advocates have not succeeded in removing the need for a black hole in this binary system. A larger sample of sources, with similar properties, may eventually lead to the general acceptance that these particular collapsed objects exist. In 1978 a second candidate, in the constellation of Scorpius, was discovered.

9.5 V 861 Sco: The second black hole?

The bright star V 861 Sco (HD 152667) is known to show binary motion with a period of 7.85 days. A suggestion by Norman Walker that it might be associated with a strong X-ray source in the constellation of Scorpius (Sco X-2) had been disproven when a more accurate position for that X-ray source had become available.

In April 1978 the bright star was studied by the Princeton ultraviolet telescope on the Copernicus satellite. This star was of interest because of the high rate at which it loses mass. The ultraviolet spectral measurements are an ideal probe of the conditions in the expanding atmospheres of such stars. As is normal, the X-ray detector of the Mullard Space Science Laboratory instrument on the satellite was also switched on and collecting data from the region containing the bright star. This part of the sky contains several weak —but well-catalogued—X-ray sources, so that it was not surprising when signals were detected by the X-ray detector. Its response covered a much larger region of the sky than the ultraviolet detector.

It was noted that the X-ray signal fell to a low level towards the end of the observation. The possibility that this behaviour may have been recurring led to a later observation, which did indeed show a similar drop in intensity. Since the interval between these low-intensity states corresponded to almost exactly four times the 7.85-day binary period of V 861 Sco, the case for an association between this star and the X-ray source was strongly reinforced. The astronomers concerned became quite excited, since like Cyg X-1 the secondary was unseen and apparently exceeded the mass limit for a neutron star. Early calculations indicated that the decreases in signals could be caused by the X-ray source being eclipsed by the bright star. The new X-ray source was given the name OAO 1653–40.

After some confusion about the phasing of the X-ray events, it became clear that the two decreases observed in the signals were *not* occurring through occultation of the X-ray source by the primary star. At these times the unseen companion was, in fact, in front of the primary star. The identification rested on the evidence of two decreases in signal at about the same phase, together with the positional coincidence which was insufficient at a value of a few square degrees. It appeared possible that substantial amounts of gaseous material were producing the eclipse-like behaviour. Several more observations of the source were made in 1978 but during July and

August the source almost disappeared below the threshold of sensitivity of the X-ray detector.

The estimates of the mass of V 861 Sco range from 20 to 50 $M_\odot$ and the mass of the unseen secondary lies in the range 7.5–13 $M_\odot$. If there were to be evidence of substantial X-ray emission it would confirm that the unseen companion is a compact object. The estimate of mass exceeds that of the limit for a neutron star by a substantial factor. This system would become the second candidate to contain a black hole.

9.6 The significance of black holes

We have described rather carefully the evidence of the existence of the black hole in the binary system containing the Cyg X-1 X-ray source. Such binary systems, if they exist in large numbers in our galaxy, will provide an excellent opportunity for us to study nearby regions where gravity dominates the other forces of nature. If we were to attempt the construction of even a mini black hole on our planet and we were to have at our disposal all the available hydrogen of the oceans, and if we could then use it most efficiently in a nuclear-fusion machine for compressing matter, we would still fall short of the requirements necessary to achieve gravitational collapse by many orders of magnitude.

Our Galaxy provides a natural laboratory for testing theories of general relativity. The instruments for the investigations of early astronomers, like Galileo's telescope, are the same as those used to observe a manœuvring enemy from a superior distance. Similarly, today we have the opportunity to turn our attention towards the stars to observe the energy released in gravitational collapse and the subsequent accretion of material by these collapsed regions of space and time. The man-made destructive big bangs are feeble compared with the energy released by nature's singularities.

We have begun to locate the first examples of black holes in our Galaxy. These tentative steps are giving us confidence that we will eventually understand the continuous production of great quantities of energy in very small regions of space. In the next chapter we take a leap to the nearby galaxies and beyond. The signals from these galaxies are more difficult to measure, but the amounts of energy released in the central regions of some of them may force us to conclude that very massive black holes lurk there, feeding on stars which have wandered into unstable orbits.

10 · X-rays from galaxies and quasars

10.1 Galactic types and distances

Up to now we have concentrated on X-ray sources within our own Galaxy, a normal galaxy in which the majority of the stars exist in spiral arms together with large quantities of gas and dust. There is also a central region, known as the nucleus, which contains a high density of stars. There is evidence from radio observations for explosive energy releases having occurred in the nucleus in the past. Knowledge of the structure of our Galaxy has come mainly from radio observations. However, it would have been very difficult to construct a coherent picture based on these data alone had it not been for our ability to see other spiral galaxies in the Universe. A typical example is given in Plate 18, where the galaxy M 51 is shown together with its companion, the irregular galaxy NGC 5195. The spiral arms and the bright central nucleus of M 51 are apparent in Plate 18. The irregular galaxy has no obvious structure and instead consists of a mass of gas and stars. The Large and Small Magellanic Clouds are the closest and best-known examples of irregular galaxies.

There are a large variety of spiral galaxies. Some have simple nuclei like the object shown in Plate 18. Others have their spiral arms joined to extended bars of gas and dust rather than to the nucleus itself. These objects are known as barred spirals. Other galaxies are elliptical in shape and have no spiral structure. It had been suggested that a sequence of increasing galactic age existed from irregulars through spirals and barred spirals to ellipticals, but this view is no longer accepted. Instead, it is believed that the different types of galaxy are distinguished by different rates of star formation and associated different rates of gas depletion. Stars are therefore formed most rapidly in irregular galaxies and least rapidly in ellipticals.

There are many other galaxies like our own, but so far X-ray emission has been detected from very few of them. Individual X-ray sources, similar to those in our own Galaxy, have been found in the Magellanic Clouds. Weak X-ray emission has been detected from the nearby spiral galaxy M 31, the Andromeda Nebula, but once again, this is probably from familiar sources like the binary-star systems and supernova remnants that have been discussed in previous chapters. At present there is great interest in the intense sources of X-ray emission that exist in the nuclei of certain active galaxies and quasars. Before discussing these, it is important to summarize the methods used to estimate the distances of extra-galactic objects from the Sun.

Distances are important in astronomy for at least two reasons. We can learn about the overall size of the Universe by establishing a distance scale for very distant objects. It is also vital to know the distances of sources if we are to determine their power output or luminosity. This can be estimated by application of the inverse-square law (see Chapter 1), according to which the apparent brightness of a source falls, in the absence of absorption, with the square of its distance.

Many different methods of measuring astronomical distances are used but a detailed discussion of them would be out of place in the present book. Instead, a brief summary will be presented.* Within the solar system, distances to planets are measured by *radar* techniques. A precisely timed pulse of microwave energy is directed towards the Moon or a nearby planet and a measurement of the time interval between the transmission and the receipt of a reflected signal is made. Knowledge of this time interval and the velocity of electromagnetic radiation allow the distance of the planet to be estimated. This method is accurate but since the size of the solar system is only 1 light day and the nearest star is 4.3 light years away, use of the radar technique is limited to the solar system.

The method of *parallax* consists of measuring the apparent change in the position of a star as seen from the Earth while the latter travels in its orbit around the Sun. The Earth travels around the Sun in an orbit of radius 1.5×10^8 km. When the star is viewed from opposite ends of an orbital diameter a knowledge of the change in the position angle, and of this baseline of 3×10^8 km, allows us to estimate the distance to the star. The parallax, is, in fact, defined as

* For a comprehensive account, refer to *Exploring the Galaxies* by Simon Mitton.

one-half of the apparent angular displacement of the star, and the distance at which the parallax is one arc second is called the *parsec* (pc); one parsec equals 3.26 light years. Although this method is also highly accurate, it cannot be used to estimate distances greater than about 300 pc (980 light years), when it becomes difficult to measure the correspondingly small angular displacements. Hence we can only measure with geometrical precision the distances of the nearest few hundred stars. Fortunately this is sufficient to permit us to proceed to the next method of distance measurement.

If we knew the intrinsic luminosity or power output of all the members of a class of stars, we could deduce the distance to any of these stars by observation of their apparent brightness and from the knowledge of their intrinsic luminosity. Thus, if we independently measure both the brightness of and the distance to a standard star, then whenever we observe another star of the same class, we can establish its distance by a measurement of its apparent brightness and an application of the inverse-square-law dependence of brightness on distance. This method of *standard candles* has many variations. It is applied not only to stars but also to novae and supernovae, to radio emission from hydrogen gas clouds and even to entire galaxies. Indeed, we can use this method with any object whose physics is well enough understood to permit a secure knowledge of its luminosity. However, it is unfortunately the case that the greater the distance involved, the greater the uncertainty in its measurement. It is only within a nearby group of galaxies known as the *local group* that we can cross-check our results by several independent methods. The largest distance within this group is around 800,000 parsec.

A class of star called *Cepheid Variables* has played an important role in distance estimation. In 1908 Henrietta Leavitt of Harvard College Observatory established a relationship between apparent brightness and period or rate of variability for the Cepheid variables in the Small Magellanic Cloud—a nearby irregular galaxy. These stars vary in brightness with an established cycle or period of between one and twenty days. The longer the period, the greater the overall brightness of the star. Harlow Shapley was able to establish the distances to similar stars within our Galaxy by trigonometrical methods, and so convert Leavitt's period–brightness law into a period–luminosity relation. Shapley unfortunately failed to recognize the existence of two different kinds of Cepheid and the resulting discrepancy in distance estimates was not corrected until the work of Walter Baade, Robert Kraft and others in the early 1960s.

However, the discovery of the period–luminosity relation opened the way for Hubble's pioneering work on the recession of the galaxies which we will now discuss.

If a star or galaxy is emitting a spectrum line and moving away from us at a velocity v, then, due to the Doppler effect, the wavelength (λ) of the line is altered by an amount $\Delta\lambda$ from the rest value where $\Delta\lambda/\lambda = v/c$ and c is the velocity of light. Since we know the value of c, a knowledge of λ and an observation of $\Delta\lambda$ will permit us to establish the recession velocity v. Hubble concluded in 1929, from observations of $\Delta\lambda$ in the spectra of a number of galaxies and from independent determinations of their distances, that the velocity of recession of each galaxy was proportional to its distance or $v = Hd$, where H is the now-famous Hubble constant. A plot of recession velocity against distance from recent work by Allan

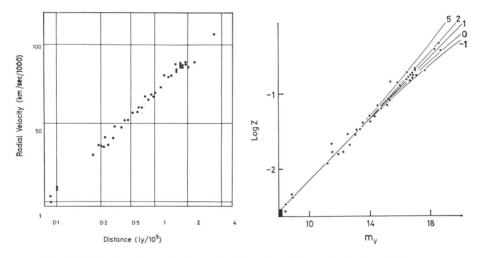

Fig. 10.1 (a) Recession velocity is plotted against distance for the brightest galaxies in each of forty-one clusters. Data were obtained by Sandage
Fig. 10.1 (b) The same data but with red shift (Z) plotted against magnitude (m_v). The solid lines are drawn for various values of q_o., the cosmological deceleration parameter

Sandage is shown in Fig. 10.1(a). The points represent data from the brightest galaxies in each of forty-two clusters (see Chapter 11). H is a velocity per unit distance and is usually expressed in kilometres per second per million parsec (km/sec/Mpc). Current estimates of its value vary between 50 and 100 km/sec/Mpc. The same data are plotted in Fig. 10.1(b) but this time as red shift (Z) against magnitude (m_v) or apparent brightness. The solid lines drawn through the points correspond to various values of q_o, the cosmological-deceleration parameter. We will return to a discussion of this figure in the next chapter when we outline the cosmological importance of X-rays from cluster sources.

We have now described methods of measuring distance that are believed to be valid even for the most distant galaxies. Knowledge of these vast distances helps us to appreciate the considerable X-ray power being radiated by a number of unusual objects, which we will now discuss.

10.2 Cen A: An unusual and energetic object

Cen A is the nearest of the radio galaxies, at a distance of 13,000,000 light years or 4 Mpc. Radio galaxies consist of large extended regions of radio emission or lobes with an active optical galaxy located between them. Cen A provides a good example of this structure as is shown in Plate 10(a). The radio intensity is displayed as a contour map in which individual contour lines enclose regions, where the radio emission exceeds a certain flux-density. The double-lobe structure, which is spread out on either side of the optical galaxy (NGC 5128), has an angular extent of around 10° on the sky. The scale mark shows that this corresponds to a distance greater than 800 kpc or 27,000,000 light years.

Very many radio galaxies exhibit this characteristic double structure, which is thought to result from explosions in the nucleus of the central active galaxy. Clouds of high-energy plasma are emitted in these explosions and the radio-emitting regions delineate the present positions of these expanding clouds. The radiation is generated by synchrotron emission (see Chapter 4) from the electrons of the plasma as they interact with the magnetic field that is carried out with the expanding plasma. The operation of the synchrotron process is evident from the polarized nature of the radio emission. The directions along which the plasma has been ejected correspond to the rotation axis of the central galaxy.

The optical galaxy itself, NGC 5128, is an elliptical galaxy. It is unusual in that its central regions are obscured by a pronounced dust lane. However, signs of more recent activity may be deduced from the presence of two small eliptical radio-regions within the optical galaxy and on opposite sides of its nucleus (see Plate 10(c)). Although the dust lane presents optical observations of the nucleus of NGC 5128, a compact non-thermal radio source has been identified with linear dimensions of less than 10 pc. Indirect evidence from the radio emission of this source suggests that its size may be even as small as 0.1 pc. Kellermann has in addition observed the nucleus at 1-mm wavelengths and has identified possible variability on a time scale of one day. If an object varies within a particular time interval, then the size of such a source must be comparable to the distance which light can travel in the same interval of time. This is because nothing can travel faster than the speed of light according to the special theory of relativity. Since the agent or disturbance responsible for the variability must be subject to this limitation, the size of the varying source must therefore be 1 light day or 0.001 pc, a distance not much larger than the size of our solar system.

Given the non-thermal nature of the radio and infrared emissions from the nuclear source, it is clear that the radiation can not arise from a single star or even from many stars. In any case the required output of energy is too large to be supplied by normal stars. It was therefore not too surprising when, in 1969, Stuart Bowyer and his colleagues at the University of California at Berkeley detected X-radiation from Cen A during a brief rocket flight. This observation was followed almost two years later by the detection of an X-ray source coincident with the galaxy NGC 5128 by the Uhuru satellite. This work specifically excluded the radio lobes as major sources of X-radiation, although they were later identified as weak X-ray sources by Lawrence using the Ariel V sky survey instrument (see Plate 10(a)).

The compact nature of the X-ray source was established with the discovery of variability by Davison, Culhane and their co-workers using the Mullard Space Science Laboratories Copernicus X-ray telescopes (see Fig. 5.4). Following this observation, Winkler and White examined Cen A data obtained with the MIT's instrument on the OSO-7 satellite. They confirmed the result obtained with the Copernicus satellite and found further evidence for variability on time scales as short as six days. More recently both the Mullard Space Science Laboratory Ariel V and OSO-8 Goddard Space Flight Center instruments have observed the nuclear source to vary in times of less than one day. Thus, the remarkable variability of the nucleus at radio and infrared wavelengths is also seen in the X-ray range. Since the original detection of variability by Copernicus, a pattern of long-term (year-to-year) variation coupled with shorter term fluctuations has emerged. It is interesting that the change in X-ray output, which occurred between 1972 and 1976 (see Fig. 5.4), is 10^{43} J or the equivalent of one supernova explosion.

Positional measurements also emphasize the location of the X-ray source in the nucleus of NGC 5128. A series of successively more severe limits on source-extent and position were established by Davison and co-workers with Copernicus, by Grindlay using instruments on the American–Netherlands satellite (ANS) and finally by Schnopper with the rotation modulation collimator instrument on the MIT SAS-3 satellite. These results are illustrated in Plate 10(c). The SAS-3 result, in particular, demonstrates that the X-rays originate from a point (less than 1 arc second) source located within 10 arc seconds of the galactic nucleus.

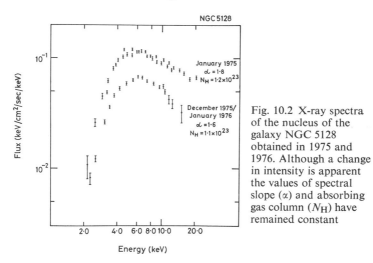

Fig. 10.2 X-ray spectra of the nucleus of the galaxy NGC 5128 obtained in 1975 and 1976. Although a change in intensity is apparent the values of spectral slope (α) and absorbing gas column (N_H) have remained constant

Measurements of the X-ray spectra of this source provide further interesting information. Two spectra were obtained one year apart by the Mullard Space Science Laboratory spectrometer on Ariel V and are shown in Fig. 10.2. Besides demonstrating the change of intensity that occurred in one year, they also show a pronounced turnover at lower X-ray energies, which is due to absorption of the X-rays by material surrounding the nucleus. Although this absorption could be caused by material anywhere along our line of sight to the source, we know from other measurements that the required quantity of material cannot exist in our own Galaxy. Thus the nucleus of NGC 5128 must be surrounded by a dense cloud of material. This will have important implications when we come to discuss models of Cen A and similar objects.

Our knowledge of this remarkable object is complemented by the extension of the X-ray spectral measurement to 5 MeV (5,000,000 electron volts), a photon energy which is well into the gamma-ray range. This work was carried out by Hall and co-workers at Rice University in Dallas, Texas who used a balloon-borne scintillation detector. They have also claimed the discovery of emission lines at 1.6 MeV and 4.5 MeV, which would have to be produced in nuclei rather than in atoms and would therefore indicate the presence of very high-energy particles indeed. In the study of X-ray emission and other high-energy phenomena, Cen A is playing a role in the Universe at large similar to that of the Crab Nebula in our own Galaxy. We will return to this source in the last section of this chapter, where we will attempt to explain the extraordinary quantity of energy released in the nuclei of this, and other, active galaxies.

10.3   Active galaxies, Seyferts and quasars

Although some traces of activity are to be found in the nuclei of normal galaxies such as our own, the quantities of energy released appear to be relatively small. In the previous section we discussed the radio, infrared and X-ray evidence for large energy releases in the nucleus of the elliptical galaxy NGC 5128. However, it is clear in Plate 10(b) that the nucleus of this galaxy is covered by a dust lane, and much of the interesting structure is therefore obscured.

Not all active galaxies suffer this obscuration. Seyfert galaxies, many of which have recently been associated with X-ray sources and which are named after their discoverer Carl Seyfert, are a class of spiral galaxies with unusual nuclei. These small, intensely bright, nuclei have optical spectra which contain broad emission lines indicating the presence of ionized elements and high-temperature gas moving at high velocity, which is not associated with individual stars. The brightness of the nuclei are such that while they can be discerned in short-exposure photographs, much longer exposures are required to see the relatively faint spiral structures. At the present time more than 120 Seyfert galaxies have been identified.

The optical spectra of Seyfert galaxies are complex but can help us to understand the nature of the Seyfert nuclei. Forbidden lines of highly ionized atoms (e.g. Fe X, Fe XIV) can be emitted only from low-density gases such as are found in the solar corona. On the other hand, allowed or permitted lines (e.g. lines emitted by hydrogen atoms) have no such limitation. Two kinds of behaviour are observed in Seyfert galaxies and this has led to a dual classification of these objects. The spectra of type-I Seyferts show very broad hydrogen lines, which indicate gas velocities in excess of 20,000 km/sec. These lines also show variability, which indicates that they originate in a compact region of less than 0.1 pc in diameter. However, the forbidden lines are narrow, indicating gas velocities less than 1,000 km/sec. They appear to come from an extended region around the nucleus of between 100–200 pc in size. In the spectra of type-II Seyferts, both hydrogen lines and the forbidden lines are narrow.

Other distinctions are apparent between the two classes of Seyferts. The nuclei of type-I Seyferts emit blue continuous spectra described by power laws indicating a non-thermal origin for the radiation. This non-thermal continuum probably extends into the infrared part of the spectrum, which contains a significant fraction of the total luminosity. However, in the case of type-II Seyferts, the infrared radiation is believed to come from heated dust grains. Although many Seyfert galaxies are relatively weak radio sources, radio emission appears to be more often associated with type-II Seyferts where extended radio-emitting regions have been resolved. Weak, unresolved and presumably non-thermal radio sources are found to be associated with type-I nuclei. We will see in the next section that the association of Seyfert galaxies with X-ray sources is so far valid only for type-I Seyferts. There is additional evidence to suggest that the X-rays are emitted from the very localized regions (<0.1 pc) where the non-thermal radiations originate.

Quasi-stellar objects—or quasars—are even more remarkable. In 1960 workers at Jodrell Bank had established that many of the newly discovered radio sources were objects of small angular diameter. This was considered interesting and led to the discovery by Sandage and Matthews that, in the case of four of these small-diameter radio sources, the only associated optical object appeared to be a star. These star-like objects had unusual spectra consisting of broad emission lines which could not be identified, combined with an excess of ultraviolet radiation. For one of these objects, 3C 48, the optical brightness was seen to vary appreciably in a matter of days, which suggested that the emitting-region could be as small as 0.001 pc. All of these observations suggested that the visible objects were unusual stars located in the Galaxy.

In 1963 two events occurred which dramatically changed this

situation. Cyril Hazard and his colleagues working with the Parkes radio telescope in Australia undertook a lunar-occultation observation of the radio source 3C 273 (recall our previous discussion of the use of the lunar-occultation technique to study the X-ray structure of the Crab Nebula in Chapter 6). In fact the use of this technique was pioneered by radio astronomers. Hazard and his co-workers demonstrated that 3C 273 consists of two components, one of which is coincident with a blue star-like object. This identification of the radio source lead Maarten Schmidt to obtain an optical spectrum of the object. He discovered a faint blue jet which extended towards the second radio component but he also noted that the spectrum of the star-like object contained the same broad unidentified emission lines that had been noted to originate from the companion of 3C 48. Schmidt decided to interpret the spectrum in terms of a red shift and was immediately successful; he found that $Z = 0.158$ permitted the emission lines to be identified with hydrogen.

Progress was rapid following this remarkable result. At the present time several hundred of these high-red-shift star-like objects are known. The history of their discovery led to their being described as Quasi-stellar Objects (QSOs) or quasars. However, they are no longer believed to be stars. If we interpret the red shift of 3C 273 ($Z = 0.158$) in accordance with the Hubble relationship (see Fig. 10.1(a)) then the distance to this source must be 480 Mpc. If this value of distance is correct, the observed brightness of the object (thirteenth magnitude) means that its real or absolute brightness must be more than one hundred times that of the brightest known galaxy. Yet we are talking about an object whose optical emission varies on a time scale of days, indicating a size of around 0.001 pc. This, then, is the central problem posed by QSOs, namely how is it possible to generate an output of energy more than one hundred times that of a galaxy, in a volume that is comparable to that of the solar system. Since examples with Z greater than 3—corresponding to distances of 4,800 Mpc—have been found, the problem is then posed in even more acute form than in the case of 3C 273.

The difficulty of the large quasar-luminosities is so severe that many people have questioned whether or not the Hubble relation applies to objects of such large red shift. Their objections are described in Simon Mitton's book *Exploring the Galaxies* but there is at present fairly general agreement that the red shifts or quasars indeed indicate large recession-velocities and therefore very large distances. In the next two sections we will discuss the X-ray observations of Seyferts and quasars and a number of possible mechanisms for energy release.

10.4   The X-ray properties of active galactic nuclei

In a previous section, we discussed the unusual radio galaxy Cen A and found it to be a variable X-ray source. Although faint X-ray emission has been detected from the extended radio lobes, a strong variable source exists in the nucleus of the galaxy. Since the optical galaxy NGC 5128 is obscured by a dust lane, it is not possible to learn anything about its appearance at visible wavelengths. However, other active galaxies have also been identified as X-ray sources, and it is the properties of these objects and of X-ray-emitting quasars that we now wish to discuss.

As remarked in the previous section, Seyfert galaxies have been identified as a class of X-ray-emitting objects. The first of these sources to be discovered, NGC 4151, was identified as an X-ray emitter by Herbert Gursky and his colleagues using the Uhuru satellite. Once again, the small size of the X-ray-emitting region has been emphasized by the discovery of variability. Several different types of variation have been observed in this source. Using the proportional counter instrument on the OSO-8 spacecraft, Richard Mushotzky and his co-workers at the Goddard Space Flight Center have found the X-ray flux in the 2–60 keV photon-energy range to vary on a time scale of a few days. Even more rapid fluctuations (time scale about 1,000 seconds) have been claimed by Harvey Tanenbaum and his colleagues who observed the source with Uhuru. If the latter observation is confirmed, it will indicate that the size of the emitting region is little more than twice the distance between the Earth and the Sun.

No change in the source spectrum was detected during the variation in source intensity observed by the Goddard group. However quite a different sort of variation had been observed earlier by Sanford and colleagues using the Ariel V X-ray spectrometer. Two spectra obtained by that instrument over an interval of more than two years are shown in Fig. 10.3. Both these spectra turn over at low energies but this turnover, which results from absorbing material, is much more pronounced in the December 1976 spectrum than in the November 1974 observation. In addition the 1976 data show a step or edge at a photon energy of 6.4 keV. This edge is due to selective absorption by iron atoms. In fact the column of absorbing material in front of the source has increased in density by about four times, and this additional material is preferentially absorbing the lower energy X-rays. The increased column of absorbing gas is probably caused by the passage of a gas cloud in front of the X-ray source. This cloud must contain as many as 2×10^{23} atoms along our line of sight. If we assume that its transverse dimension is just sufficient to cover the X-ray source, then its velocity is greater than 2,000 km/sec, a value consistent with that deduced from the widths of the broad emission lines seen in the optical spectra of Seyfert galaxies. Hence, the X-ray observations can tell us about the size of the nucleus and the nature of the regions surrounding it.

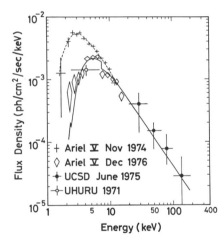

Fig. 10.3 X-ray spectra of the nucleus of the galaxy NGC 4151. Although the spectral slope remains the same, a considerable change has taken place in the size of the absorbing gas column

Following the discovery of NGC 4151 by Uhuru, the Ariel V sky survey detected X-rays from several more Seyfert galaxies. These were confirmed in a further edition of the Uhuru catalogue, which also contained a number of additional identifications while observations with MIT's SAS-3 satellite have added yet more examples. However, mainly due to the Ariel V sky survey, there are now twenty-five reliable identifications of X-ray sources with Seyfert galaxies. In fact, it is interesting to notice that four of these identifications came about through searches of X-ray-source error boxes by optical astronomers, which led to the discovery of new Seyfert galaxies in the southern skies.

With a larger sample of X-ray Seyferts now available, it is possible to compare the properties of these objects at several different wavelengths. The X-ray luminosity correlates well with the optical and infrared continuum radiation and with the power radiated in the Hα emission. However, it does not correlate with radio power or with the intensity of the optical forbidden lines. In addition, all of the X-ray-emitting Seyferts discovered so far are type-I Seyferts. These features of the X-ray sources in Seyfert galaxies indicate that the X-rays are generated in the nuclear regions and are therefore intimately connected with the mechanism of energy release.

The nuclear origin of the X-rays is also supported by the observation of variability. We have discussed this in some detail in the case of NGC 4151 where variability on a time scale of days and possibly of minutes has been detected. Observations by Ariel V and other spacecraft have indicated that several other X-ray Seyferts are variable. The intensities of the sources are low and it is not yet possible to establish whether changes in X-ray flux are due to changes in source emission or changes in the column of gas in front of the source. However, some spectral data has been obtained by the Goddard Space Flight Center instrument on HEAO-1, which suggests that the large columns of gas observed in front of the nuclei of Cen A and NGC 4151 may be the exception rather than the rule. Spectra of about five active galaxies obtained by the Goddard group show no evidence for the presence of X-ray absorption.

With the number of X-ray-emitting Seyferts now available, it is possible to make a preliminary estimate of the X-ray luminosity function of Seyfert galaxies. The luminosity function is simply a statement of how many sources there are of a given X-ray luminosity per unit volume of the Universe. A luminosity function for Seyfert galaxies has been derived by Elvis and colleagues, of Leicester University, using the Ariel V sky survey data. They assumed no cosmological evolution of these sources (i.e. X-ray emission from Seyferts has always been the same and was not significantly greater or less at earlier times). With this condition, they established that Seyfert galaxies can not contribute more than 6 per cent of the observed diffuse X-ray background flux (see Chapter 12). If all X-ray-active galaxies and quasars are included in the luminosity function, their total contribution to the diffuse background still remain less than 15 per cent.

Quasars are now established (from the results of the Einstein X-ray Observatory) as a class of X-ray sources. For many years the nearest of these objects, 3C 273, remained the only known X-ray-emitting quasar but the number of X-ray quasars is now increasing rapidly. If we examine the X-ray luminosities of quasars and Seyferts, it is interesting to note that the luminosity of the brightest Seyfert is just a factor two below the luminosity of the weakest quasar. This lends support to the idea that quasars, Seyferts and other active galaxies may well be powered by a similar mechanism in their compact nuclear regions.

10.5   Active galactic nuclei: giant pulsars or giant black holes

Although X-ray studies of the nuclei of active galaxies have only just begun, there is reason to believe that observations in the X-ray range may at last enable us to come to grips with the problem posed by these objects. The essential difficulty, put in its most acute form by the quasars, is how to explain the release of energy at more than 10^{39} J/sec in volumes that are of order 0.003 light years (0.001 pc) or less in diameter. Associated extended radio structures require that the energy source have a lifetime of 10^5–10^6 years, and thus the total energy requirement could be as much as 10^{53} J.

We have discussed the observational situation in the preceding sections and have seen that the cores of active galaxies emit a continuous spectrum of radiation from radio wavelengths through the infrared and optical bands to the X-ray range. This emission appears to be non-thermal although it could be produced by hot gas with a range of different temperatures. It varies in times that may be as short as days and is therefore emitted from small regions. This is further emphasized by the detection of variable X-ray absorption probably caused by the motion of clouds or filaments in front of the nucleus. As a prelude to understanding the energy release mechanism, it is necessary first to identify the X-ray production mechanism and here two possibilities exist—namely Compton-synchrotron radiation or Bremsstrahlung from hot gas being accreted by a black hole.

We have already discussed the synchrotron process, by which electrons in spiralling around a magnetic field emit electromagnetic radiation (see Fig. 4.3). It is believed that the non-thermal radio, infrared and optical emissions from the active nuclei are generated in this way. The Compton effect, in which photons collide with electrons and give up energy to them, was discussed in Chapter 2 (Section 2.3).

It is also possible for the opposite or 'inverse Compton effect' to occur. In this process, a low-energy (radio or infrared) photon can interact with a very high-energy electron and gain sufficient energy from the collision to be transformed into an X-ray. Because of the extremely compact nature of the active galactic nuclei, it is possible for radio and infrared photons, produced as synchrotron radiation from high-energy electrons, to interact again with these same electrons by the inverse Compton process. We will discuss this mechanism further in the next chapter (Section 11.2). Following such interactions, a strong flux of X-radiation would result. This radiation is called Compton-synchrotron radiation and could explain the observed X-ray emission. If this view is correct, all of the radiation produced in the compact nuclei of active galaxies is due to the presence of very high-energy electrons and strong magnetic fields, and a mechanism must be found for supplying the required energy in these forms.

One way to accelerate electrons to very high energies and to produce the necessary magnetic fields has been proposed by Philip Morrison of MIT and a number of his collaborators. It involves scaling-up the pulsar phenomenon that was described in Section 6.2. As we discovered in Chapter 6, pulsars are very efficient electron accelerators. If we assume a conversion efficiency of 10 per cent, then the energy sources required in the nuclei of active galaxies would—in the most extreme case of quasars—be large bodies of close to 10^8 M$_\odot$ with magnetic fields of 10^5–10^6 Gauss and rotating with periods of about around five years. If such a body were around 0.001 pc (0.003 light years) in diameter, as is required by the observed variability, its density would have the quite modest value of 10^{-6}–10^{-7} g/cm^3. There is as yet no evidence for the existence of such a large rotating body in the nucleus of an active galaxy. However, the operation of the synchrotron and inverse Compton processes would provide related radio and X-ray variability, which should soon be detected if it is present. In addition it should eventually be possible to find variability on the time scale of the large body's rotation period, if this is indeed the energy-supply mechanism that is operating.

A different approach to the problem of active-galaxy energy sources has been adopted by Martin Rees and his colleagues at the Cambridge Institute of Astronomy. Once again the scaling-up of a system from our Galaxy is used. In Chapter 9 we described the generation of X-rays by the accretion of gas on to a solar-mass black hole. The gas is supplied by a normal star that finds itself in a binary system with the black hole. In falling towards the hole, the gas from the companion star is heated to a very high temperature due to the intense gravitational field. In addition, because it possesses angular momentum, the in-falling material cannot fall straight into the black hole but instead forms an accretion disc around it. In attempting to explain the source of energy in an active galaxy, Rees and his co-workers propose that a black hole of 10^7 M$_\odot$ is located in the nucleus. Material is supplied to the hole in the form of gas from the galaxy or even by the break-up of stars caught in the gravitational field of the black hole. If the gas were heated to a temperature of 10^9 K, the observed X- and gamma-ray emissions could be explained. The X-ray variability would be due to variations in the supply of accreted material. Shockwaves set up in the in-falling gas would accelerate electrons to very high energies and would therefore provide the observed radio, infrared and optical-synchrotron emission.

Once again there is no firm evidence for the existence of black holes in the nuclei of active galaxies and quasars. However, the detailed study of variations in the X-ray intensity and spectrum could provide the necessary confirmation of this hypothesis. It is significant that observations of X-ray sources in our own Galaxy have led to the formulation of models, which appear at last to offer plausible explanations for the problem of energy release in active galaxies, one of the outstanding astrophysical problem of the present time.

11 · Clusters of galaxies

11.1 The nature of clusters of galaxies

We now know from the work of Abell, Zwicky and others that almost all galaxies occur in aggregates that range from pairs through small groups of a few galaxies to giant clusters which contain hundreds or even thousands of galaxies. The larger the number of member galaxies, the *richer* the cluster is said to be. These clusters are distributed uniformly in the sky and so it is now clear that clusters of galaxies must represent the unit of matter, or building brick, from which the Universe is assembled.

The size of a cluster can be established only if we know its distance from the Sun. As discussed in the last chapter, this is determined from the average recession velocity of galaxies in the cluster—by measuring the red shifts exhibited by the lines in the visible spectra of the galaxies. The Hubble relation then allows the distance of the cluster to be estimated. When the distance has been measured, it is possible to determine the linear size, provided that the extent of the cluster can be adequately defined. Distances of more than 3×10^9 light years(1,000 Mpc) have been measured for some clusters. The central core of a cluster, the region where most of the galaxies are found, is generally around 1,000,000 light years (0.3 Mpc) in diameter while the associated halo of gas may have a diameter of about twenty times this value.

Clusters appear in two main categories, namely regular and irregular clusters. The regular clusters have a marked central concentration and a large number of member galaxies. The irregular clusters, which are also sometimes called open clusters, have very little symmetry and no marked central concentration. The regular clusters frequently have giant cD-type galaxies at their centres. These are especially large elliptical galaxies that are often also strong radio sources. In general the regular clusters contain mainly elliptical galaxies while the irregular clusters contain galaxies of all types but mainly spiral and irregular galaxies.

Clusters were probably formed from the primordial gas in the Universe, which expanded following the big bang. It is likely that the largest clusters probably formed first and used up large fractions of the gas. More recent clusters are generally smaller. However, clusters should contain some material left over after the galaxies have been formed and we shall see that the X-ray observations support this idea.

We remarked above that clusters often contain strong radio sources. In fact, a large fraction of all strong radio galaxies are located in rich clusters. This is because bright elliptical galaxies, a major class of strong radio galaxies, are themselves more concentrated in rich clusters. In particular the massive cD galaxies found at the centres of regular clusters are frequently bright radio sources. The probability that a cluster will contain a radio galaxy thus appears related more to the regularity of the cluster than to its richness. In addition the radio galaxies found in clusters appear to have steeper radio spectra than those outside. Some radio galaxies exhibit a characteristic head-and-tail-extended structure, which is seen only if the galaxy is a member of a cluster. Although difficult to observe, large low-frequency radio sources are occasionally found to be associated with clusters of galaxies. These extended regions are much larger than a single radio galaxy and are sometimes comparable in size to the cluster itself. We will return to a discussion of cluster radio properties after we have introduced the idea that clusters are extended X-ray sources.

Before leaving a general discussion of clusters, it is of interest to inquire why these assemblies of galaxies should still exist in the observable Universe. In pursuing this question we encounter an interesting difficulty known as the missing-mass problem. A number of methods exist for determining the masses of individual galaxies. We can observe the rotation of our own and of the nearby galaxies and compute their masses with the help of Kepler's third law. If a pair of galaxies exist in close proximity, we can estimate their masses by observing their gravitational influence on each other. In addition radio observations of the hydrogen line at a wavelength of 21 cm can allow us to estimate the mass of gas in a galaxy. By undertaking observations of this kind for many different types of galaxy and by observing the total light output of such galaxies, we can compile mass to light ratios for the various galaxy types. If we now wish to determine the mass of material in a cluster of galaxies, we first determine the masses of the individual members using the methods just described. We can then deduce the total cluster mass by adding up the masses of the individual members.

There is yet another way to estimate the mass of the cluster as a whole. All of the galaxies in the cluster are moving, both with respect to us and to each other. As remarked at the beginning of this section, the average velocity represents the recession velocity of the cluster in the expansion of the Universe. The mean of the differences between the velocities of the individual galaxies and this average is called the velocity dispersion. If we assume that the individual galaxies have been accelerated to their observed speed by the gravitational attraction of the mass of the cluster as a whole, it is possible, by means of a theorem known as the virial theorem, to estimate this total mass from the velocity dispersion of the individual galaxies. The result for all clusters obtained from this method is the so-called virial mass. This can be up to ten times larger than the mass obtained by summing the contribution of the individual galaxies. This is the missing-mass problem referred to above. Since the clusters appear to hold together, the gravitational cluster mass of the cluster obtained from the virial theorem must be correct and the missing mass must exist in an as yet undetected form.

11.2 Clusters of galaxies as extended X-ray sources

It will have become clear from the previous section that clusters, as the building blocks of the universe, are of major importance in astronomy. Unfortunately, optical and radio observations tell us little about their collective properties; it is not yet possible to observe general optical emission and extended radio sources of the size of the clusters are barely detectable in a very few cases. However, observations in the X-ray wavelength have proved to be of great value.

In the era of sounding-rocket observations, Friedman's group at the NRL detected X-rays from the well-known active galaxies M87 and NGC 1275, which are in the Virgo and Perseus clusters respectively. However, the true nature of these X-ray sources became clearer after the successful launching of the Uhuru satellite late in 1970. The most important discovery made by Uhuru in the observation of extra-galactic X-ray sources was that extended regions of X-ray emission were associated with a number of clusters of galaxies including those in Virgo and Perseus. Although the field of view of the Uhuru detector was as large as 30 arc minutes in the scanning direction, it was still possible for Uhuru to demonstrate that the angular extent of a number of clusters was in the range 30 arc minutes to 1 degree—values comparable with the overall sizes of the optical clusters. Linear dimensions therefore ranged from one to several million parsecs. The X-ray power emitted from these sources was substantial and for the first four X-ray clusters discovered lay in the range 2.10^{36}–10^{38} W. The large angular sizes and power outputs led the Uhuru group to conclude that the emission from the clusters is diffuse, rather than from a number of individual X-ray sources.

The nature and origin of the X-ray emission became a matter of intense speculation following the discovery of these extended objects. The existence of extended low-frequency radio sources, associated with several clusters, led to suggestions that the inverse Compton mechanism is the origin of the X-radiation (recall the description of the Compton effect presented in Chapter 2 and Fig. 2.1(c)). Here a photon interacts with an electron, gives up some of its energy to that electron and leaves the interaction site in a different direction. Although the cross-section for the reverse process (in which a photon gains energy from an electron), is small, it is possible for it to occur in the vast volume of the Universe. The process proceeds in the following way: if a low-energy photon encounters a very high-energy electron it can gain much of that electron's energy, and emerge transformed into an X-ray from the interaction.

A ready supply of low-energy photons for this process exists throughout the Universe in the form of the 2.7 K microwave background radiation. This radiation is believed to be the residue of the radiation that was produced in the 'big bang' creation of the Universe. The radiation has cooled with the passage of time to its present temperature. It has in addition participated in the universal expansion that followed the 'big bang' so that it now permeates the entire Universe.

We have seen in the last sections that clusters contain many radio galaxies. Since the radio emission is due to synchrotron radiation from very high-energy electrons, it is probable that explosions and other forms of activity in these galaxies will eject large numbers of electrons into the space surrounding the galaxies. The sum of the ejected electrons from a number of galaxies could provide a large pool of very high-energy electrons in a cluster. The photons of the

Universe's microwave radiation would interact with these electrons by the inverse Compton process and produce a flux of X-rays from a cluster containing such high-energy electrons. Thus this mechanism offers one possible explanation for the existence of extended X-ray sources associated with clusters of galaxies.

It is also possible that the clusters are filled with high-temperature gas. In Chapters 3, 4 and 6 we discussed the gases that have temperatures up to 6×10^6 K and are responsible for the observed X-rays in both the Sun and the supernova remnants. The X-ray spectra, observed in the case of clusters, require gas temperatures in the range 2×10^7 to 2×10^8 K, if they are to be explained by the presence of hot plasma. The discovery of a large amount of hot gas distributed throughout the Universe in clusters could have important implications both for the clusters themselves and the Universe as a whole. We will return to the question of the origin of the observed X-rays later.

Following the discovery of cluster X-ray sources by the Uhuru satellite, the subject developed rapidly. The third Uhuru catalogue of X-ray sources lists about twenty possible identifications of known clusters of galaxies with X-ray sources. Unfortunately, identification is rendered difficult by the large positional uncertainty associated with weak X-ray sources. This arises from the rather large field of view of the collimated X-ray detector (see Fig. 2.3). One of the Uhuru detectors had a field of view of 0.5° by 5° in angular size. With this sort of detector the area of uncertainty (error box) associated with a weak X-ray source is at least one square degree.

Pounds and his colleagues at the University of Leicester carried out an X-ray sky survey using proportional counter detectors on the UK Ariel V satellite. By arranging the manner in which the satellite scanned the sky so that the majority of sources were scanned in a large number of different directions, they were able to reduce the uncertainty in position of many X-ray sources. Thus the majority of the sources were located in error boxes of 0.5 square degrees or less in size and this permitted the total number of X-ray-source identifications with clusters of galaxies to be increased to around fifty.

The existence of such a large (for X-ray astronomy!) sample of objects has permitted a number of interesting discoveries to be made about cluster X-ray emission. If the distance of a cluster is estimated from its average recession velocity, as discussed in the previous section, it is possible to calculate the intrinsic X-ray output power or luminosity of the cluster. The sample of fifty clusters may, with suitable selection to assure its uniformity and completion, be used to establish the number of X-ray clusters in each range of luminosity. This number increases rapidly with decreasing luminosity. Although the available number of clusters is still small, it is possible to infer the total number of X-ray clusters in the Universe from this relationship. The number equals the total number of clusters found from optical observations. We may therefore conclude that all clusters of galaxies emit X-rays at some power level greater than 10^{36} W.

The existence of the relationship between number of clusters and X-ray luminosity* described above allows us to calculate the total X-ray-output power from all clusters of galaxies in the 2–10 keV photon-energy range. This quantity may be compared with the observed flux of diffuse background X-rays (see Chapter 12). It is found that X-ray-emitting clusters as a class of objects can only account for around 15 per cent of the diffuse flux.

Returning to a discussion of individual clusters, we previously defined richness as a quantity that signifies the number of galaxies in a cluster; the richer the cluster, the greater the number of galaxies it contains. The X-ray luminosity is found to be strongly correlated with cluster richness, with the richer clusters producing more luminous X-ray sources. In addition to the richness of a cluster one can, as mentioned in the previous section, classify the distribution of galaxies in clusters as being regular (centrally condensed), or irregular. It is found that the X-ray luminosity of regular clusters is much higher than that of irregular clusters.

In the previous section we discussed regular or centrally condensed clusters, particularly those with massive cD galaxies, which often contain strong radio sources with steep radio spectra. In view of the comments in the previous paragraph, this point also may imply an association between strong steep-spectrum radio sources and high X-ray luminosity. The reason for this relationship is not yet understood, as will become clear in the next section in which we will discuss the question of how the diffuse cluster X-ray emission is produced.

Many of the points made in this section are well summarized by the data presented in Plate 15 which shows both an X-ray map and a radio map of the Perseus cluster of galaxies. X-ray mapping of the central region of the cluster, including the active galaxy NGC 1275, was first performed by Fabian and co-workers using the Mullard

---

* This relationship is known as a luminosity function.

Space Science Laboratory X-ray telescopes on the Copernicus satellite. This instrument had sufficient sensitivity to map only the region around the active galaxy. The resulting 0.5–1.5-keV map is shown as a set of contour lines in Plate 15(c). This Copernicus contour map is superimposed on a map of the entire cluster obtained more recently by Gorenstein and his colleagues of the Harvard Centre for Astrophysics, who used an imaging X-ray telescope on a sounding rocket. In addition to the same bright region around NGC 1275 that is seen by Copernicus, Gorenstein's map clearly demonstrates the extended nature of the X-ray source associated with the cluster. Interestingly, no other individual galaxies are visible in X-rays at a level of only a few per cent of the NGC 1275 signal. However, many other massive galaxies exist in the cluster and some of these are prominent radio sources, as may be seen in Plate 15(b). The role of NGC 1275 as an X-ray source is however not completely clear. The enhanced X-ray signal may be from a brighter region near the centre of the cluster or it may be originating from the galaxy itself. X-ray maps with better resolution are needed to decide this question. Plate 15(b) also shows the contours of extended radio emission discovered by Ryle and Windram. The possible existence of extended low-frequency radio sources in this and some other clusters has led people to speculate that the inverse Compton effect may play a role in X-ray generation but we shall see in the next section that this is no longer believed to be the case. Notice also the striking head-and-tail features associated with two of the radio galaxies (NGC 1265 and IC 310) in the cluster. The inset in Plate 15(b) contains a more detailed radio map of NGC 1275 taken from the work of Miley and Perola. There is some similarity between the radio and X-ray contour maps of NGC 1275 but this could be coincidental.

11.3 The production of X-rays in the space between galaxies

We outlined, in the last section, two possible mechanisms for X-ray emission from clusters of galaxies—namely the inverse Compton process and thermal Bremsstrahlung from a hot gas. We can now examine the evidence to help us choose between these models.

A study of source spectra provides the most immediate way of identifying the production mechanism but there is a difficulty. The proportional counter detector, which has up to now been the mainstay of X-ray astronomy, has rather limited energy resolution (see

Chapter 2). While it is possible, in principle, to observe the difference between the power-law spectrum arising from the inverse Compton process and the exponential (slope steepening with increasing photon energy) spectrum that would characterize thermal continuous emission, it has not proved possible to do so in practice. This is because of inadequate energy resolution and limited energy range (1–10 keV). A decade in photon energy provides an insufficient range for the difference in spectral slope to become apparent.

Lewin and his colleagues at MIT undertook an observation of the Perseus cluster, which attempted to overcome the second of these difficulties. They flew a balloon, and used a scintillation detector capable of detecting photons at energies in excess of 100 keV. X-rays of this energy were not detected from the Perseus cluster at the level predicted for an inverse Compton power-law spectrum. The absence of a signal at high energies suggests that the flux falls steeply with energy, as expected in the spectrum from a high-temperature plasma.

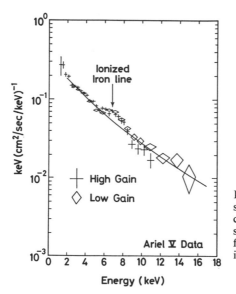

Fig. 11.1 The X-ray spectrum of the Perseus cluster X-ray source showing the emission feature due to highly ionized iron

The presence of high-temperature gas in a cluster was finally established by Culhane and co-workers, who detected an emission feature due to highly ionized iron in the spectrum of the Perseus cluster. This result is shown in Fig. 11.1 where the feature at 6.7 keV, due to line emission from Fe XXV, is clearly visible. The energy resolution of the Mullard Space Science Laboratory Ariel V proportional counter used in this observation is just sufficient to permit the detection of an isolated feature at this energy. This discovery was later confirmed by Serlemitsos and co-workers at the Goddard Space Flight Center in the USA. These workers found similar features in the spectra of the Virgo and Coma clusters. Mitchell and Culhane detected the feature in the spectrum of the Centaurus cluster and confirmed the Coma observations. Thus the presence of high-temperature plasma ($T \sim 20\text{--}200 \times 10^6$ K) in these four clusters was firmly established.

A number of the characteristics of clusters can be explained if the presence of hot gas is assumed. First, the hot gas provides a natural explanation for the appearance of head–tail galaxies (see Plate 18). The radio emission from these objects is believed to be from synchrotron radiation produced by relativistic electrons emitted by the galaxies. However, the pressure of the surrounding hot gas prevents the electrons from expanding freely into the cluster. They are, instead, confined to the immediate neighbourhood of the galaxy. The characteristic head-and-tail appearance of the radio emission is caused by the motion of the individual galaxies through the hot gas with the head being compressed and the tail extended due to the motion.

Yet another piece of evidence in favour of the existence of hot plasma in clusters comes from studies of the 2.7 K microwave background radiation, discussed above in connection with the inverse Compton effect. The microwave radiation is universal and appears to be isotropic, that is it appears to have the same intensity in all directions. Two Russian astronomers, Sunyaev and Zel'dovich, predicted that the scattering by hot electrons of the microwave background radiation, as it passed through the clusters, would alter the spectrum of the radiation so as to make it appear that a slight change in temperature had occurred. Gull and Northover, from the Mullard Radio Astronomy Observatory at Cambridge, have reported a detection of this effect as have Lake and Partridge from the University of California at Berkeley.

Returning to the X-ray observations, we find additional evidence for the existence of hot gas in clusters. If we examine the distribution of types of galaxies in clusters an interesting fact emerges. The irregular clusters generally give rise to lower X-ray fluxes and temperatures when compared with regular and more centrally condensed clusters. This implies that a smaller quantity of hot plasma exists in the irregular clusters. A large fraction of the galaxies (about 70 per cent) in irregular clusters are also spirals. For regular clusters the fraction of spirals is very much less. There is a marked and progressive decline in the fraction of spirals correlated with the increasing cluster X-ray temperature and luminosity. Spiral galaxies are similar to our Galaxy and are known to contain large amounts of dust and gas in their spiral arms. The spiral galaxies loose their gas in collisions with the hot gas as they move through the cluster and thus the spiral structure gradually disappears. This process is known as spiral stripping. The strong inverse correlation between the fraction of spirals and the X-ray temperature and luminosity thus provides further evidence of the presence of hot plasma in the clusters.

Before considering the origin of the hot gas and how it is heated, we must refer again to the iron-feature observation. The intensity of the feature can help to establish the iron abundance or quantity of iron present in the cluster. A knowledge of the abundance is, as we will see, of crucial importance in establishing the origin of the hot intra-cluster gas. The iron feature is in fact a blend of several lines from the ions Fe XXV and Fe XXVI. These individual lines cannot be resolved by the proportional counter. However, the temperature may be estimated from the shape of the continuous spectrum in Fig. 11.1. The feature intensity may be calculated from a knowledge of the temperature and of the quantity of hot gas present. However, the intensity of the feature also depends on the abundance of iron. For an unusually large quantity of iron, the feature would appear very strongly above the continuum. In the case of an exceptionally low abundance, iron emission would not have been detected by the proportional counter. In fact, the observed strength of the iron feature in Fig. 11.1 indicates that the abundance is approximately half that observed in the rest of the stars and galaxies that make up the Universe. The uncertainty in the observation is such that the cluster iron abundance in the cluster could range up to this universal or cosmic value.

We can now consider the question of the origin of the intra-cluster gas and how it is heated. It has been felt for many years that there ought to be an intergalactic medium. Just as the stars in a galaxy exist in a medium of gas and dust from which they were formed, so the galaxies should exist in a universe of the gas from which they coalesced. However, attempts to identify this gas, the so-called intergalactic medium, have so far proved unsuccessful. Nevertheless, if such a medium exists, it could provide a very large fraction of the mass in the Universe. While there is now general agreement that the Universe originated in a singular event or big bang and that it is currently expanding, its evolution remains far from clear. Depending upon the actual mass involved, the Universe could continue to expand, or it could contract again to a singularity. A reliable estimate of the mass of the Universe is required to differentiate between these two possibilities.

If an intergalactic medium exists, it could fall into the gravitational potential wells of clusters, and acquire kinetic energy in the process. Gull and Northover at Cambridge have proposed such a model in which the gas falls into the potential wells of clusters and is heated to a maximum temperature of around 10^8 K at the centre. Further out in the cluster the temperature may be less, but the gas density everywhere is so low that, even in the 10^{10}-year estimated age of the Universe, the particles would not had time to mix to produce a uniform temperature. Observations of the spectra should be able to differentiate between a single temperature and a multi-temperature gas, but this cannot be achieved with the existing data.

The detection of the emission feature from iron leads to a more serious difficulty for this model. A universal medium left over after the formation of the galaxies would not be expected to contain any iron. Iron, in common with other heavy elements, can only be synthesized in stellar interiors. No elements heavier than helium were created in the big bang. If the hot intra-cluster gas is produced by a universal medium falling into clusters, the iron feature, visible in Fig. 11.1, should not be present in the spectrum. The existence of such a feature suggests that the hot gas cannot be entirely made of material left over from the big bang. Material may have been ejected from the member galaxies of the cluster or perhaps enriched in some other way by iron from the galaxies; however, much work is needed to resolve these uncertainties.

11.4 Clusters of galaxies and cosmology

We have seen in the previous section that the extended X-ray sources, associated with clusters of galaxies, can be explained by the presence of high-temperature plasma. We also discussed briefly the question of the total quantity of matter in the Universe. In Section 11.1, we described the problem of missing mass in clusters—the puzzling anomaly whereby the sum of the masses of individual galaxies is insufficient to bind the clusters gravitationally. The discovery of a large quantity of hot plasma in clusters is relevant to both these questions. Unfortunately, the existence of this 'new' material in the Universe does not resolve either of these difficulties. Although it depends on the model for its origin, the hot gas can—at most— supply only 20 per cent of the mass needed to bind the individual clusters and only 2 per cent of the mass required to 'close' the Universe, or stop its present expansion and start it contracting again.

Thus, the discovery of hot gas in clusters does not lead to a significant revision of the overall mass of the Universe, nor yet to a complete solution of the missing-mass problem. The properties of cluster X-ray sources do, however, show promise of enlarging our understanding of the Universe in other ways.

Our traditional understanding of the Hubble relationship between recession velocity (as implied by measured red shift), and distance, is that the Universe is expanding. We have already discussed the microwave background radiation whose most likely origin was in the big bang, which gave rise to the Universe as we now see it. Given our belief in this overall picture of expansion and singular origin, we can probe the details by examining objects at large values of red shift (Z). For further details of cosmological models readers are referred to the excellent books by Dennis Sciama and Michael Rowan-Robinson.

Since we have only one universe to study, we cannot proceed as we would in the study of objects of which there are many examples. Instead we begin with a fundamental assumption, known as the cosmological principle, which states that the Universe as seen by fundamental observers is everywhere homogeneous and isotropic. A fundamental observer is one who shares in the general expansion of the Universe, unperturbed by local motions, such as that of the Earth around the Sun. The Universe is homogeneous because every observer will deduce the same picture of the Universe as every other observer. It is isotropic because it appears the same, independent of the direction in which it is viewed.

The scale of the Universe is vast, in both distance and in time. Since we can observe the Universe only by means of electromagnetic radiation, which itself travels at a finite velocity of 299,800 km/sec, we find ourselves studying the distant parts of the Universe by means of light which was emitted more than 10^{10} years ago. We must therefore think not of distance in three dimensions as we would on Earth but rather of events, such as the emission of a burst of light or photons, which occur in a four-dimensional 'space', which includes distance and time.

The mathematical treatment of this Universe is contained in Einstein's general theory of relativity, which is beyond the scope of this book. This theory includes the effects of the gravitational force in the Universe as a curvature of this four-dimensional space. In cosmology we are concerned with the gravitational force of all the mass in the Universe. Given our belief that the Universe is at present expanding, it is of fundamental importance to know if this expansion will continue for ever, will just stop, or will stop and reverse, leading to the ultimate collapse of the Universe into another singularity. We have already discussed the contribution of the hot intra-cluster gas to the mass of the Universe and found that, as observed in individual clusters, it is insufficient to check the universal expansion. However, the theory of general relativity makes predictions about the observable nature (e.g. the size and luminosity) of objects that were created early in the life of the Universe and it is the nature of clusters, in this sense, that we must now examine.

In Chapter 10 we discussed the Hubble relationships of red shift with distance (Fig. 10.1(a)) and magnitude (Fig. 10.1(b). In the latter figure, a variety of theoretical lines have been calculated for different values of q. This parameter is known as the deceleration parameter and its values specify different curvatures of space-time corresponding to different densities of matter in the Universe. For example, if $q > 0.5$, the Universe will expand to some maximum size and will then collapse again. For $q <$ less than 0.5, it will expand for ever and the galaxies would still be separating when infinitely far apart. It is clear that we cannot establish the value of q with any certainty from the data in Fig. 10.1. This is because of the wide range of luminosity exhibited by the objects and because of the evolution of their properties with time. In other words galaxies may have been a great deal more luminous in the past, whereas the theoretical lines drawn in Fig. 10.1(b) are based on our present understanding of such objects.

X-ray-emitting clusters may provide a way out of this dilemma. One consequence of the general theory of relativity is that the angular diameter of an object changes with red shift in a way which depends on the parameter q. This type of change, for individual galaxies, is illustrated in Fig. 11.2. But galaxies are complex assemblies of stars whose evolution is difficult to understand. Now that we know of the existence of high-temperature gas in clusters it will be possible to make observations of the angular diameter for X-ray clusters of varying red shift. Before we can do this we must fully understand the origin and heating mechanism for the gas and how the X-ray source changes with time. There are grounds for optimism in this approach because the physics of a hot-gas cloud is much simpler than the physics of a galaxy. It is hoped that the X-ray telescopes on the HEAO-2 spacecraft will enable us to discuss whether q is greater than or equal to 0.5 or much smaller. The present range of uncertainty in our knowledge of q lies between -1 and 2.

It is possible to estimate values of q from the X-ray luminosity of clusters in a similar manner to that shown for galaxies in Fig. 10.1(b). However, the difficulties involved with the evolution of these sources are more serious in this case. The X-ray angular diameter will evolve, together with the gas density and temperature, but even in this case the physics used is simpler than in the case of entire galaxies.

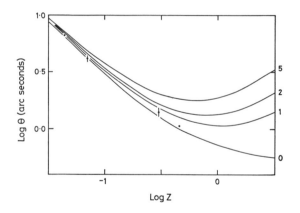

Fig. 11.2 The angular diameter (θ) of a galaxy plotted against red shift (Z). The different solid lines are for various values of the deceleration parameter q_0

12 · The observatories of the future

The first two decades of space astronomy revealed then that X-radiation is emitted, and can be detected, both from the nearest stars and from the most distant regions of the Universe. For those of us who have been privileged to play a part in the exploration of the Universe through this new window, the experience has been rewarding. In our final chapter we shall attempt to project how the new astronomy can be conducted—what kinds of observatories are needed and how their organization can advance progress over the broad frontiers of astronomy. We begin by recalling the developments of the two decades since the late 1950s, which have led to our present knowledge of the X-ray sky, and by considering some of the unanswered questions which continue to stimulate further exploration. First we discuss the problem of explaining the origins of the X-ray background—a subject which may yet provide the most significant discoveries in X-ray astronomy.

12.1 The X-ray background

After almost twenty years of investigation the X-ray background remains an intriguing problem. This background of diffuse radiation was discovered in the epoch-making rocket experiment that led to the discovery of the first source of X-radiation outside the solar system. Giacconi and his colleagues reported a diffuse background of X-radiation of 1.7 photons/cm² arriving from a solid angle of one steradian every second. The average energy of the radiation detected in their experiment was about 4 keV. Subsequent rocket and satellite measurements confirmed the existence of this diffuse X-ray background and its spectrum is reasonably well defined to very high energies of the gamma-ray region. Fig. 12.1 shows how the intensity of the diffuse background varies with energy.

The importance of the discovery of isotropic X-radiation was realized almost immediately. At energies greater than a few keV there could be no doubt that the origin of the radiation must be at great distances, well outside our own Galaxy. The strength of the observed flux of X-rays was itself sufficient to eliminate one cosmological model of the Universe. In one version of the steady-state theory it was proposed that matter was continuously created in the form of neutrons. As these neutrons are unstable they would be sufficient to heat the medium between the galaxies to a temperature of 10^9 K. This would lead to the emission of X-rays by the thermal Bremsstrahlung process with an intensity about five times the observed value of the diffuse X-ray background.

We have yet to establish how the diffuse background will aid our understanding of the large-scale structure of the Universe; however, the strength of the radiation in the energy range between 1 and 100 keV is remarkable. It greatly exceeds the intensity of radiation emitted by our Galaxy at these energies. Only in one other region of the entire electro-magnetic spectrum is a similar result found—the microwave background radiation arising from the initial big bang. The black body radiation at a temperature of 2.7 K contributes a density of radiation of about 0.3 eV/cm³. Although the X-ray background has an energy density very much smaller at about 10^{-4} eV/cm³, this density is nevertheless about ten times the energy density of X-radiation from our own Galaxy. By comparison, in most spectral regions, the starlight from the Milky Way is about one hundred times more intense than the optical flux from the distant galaxies.

The excess of radiation from regions external to our Galaxy leads to the first important conclusion about the origin of the X-ray background: ordinary galaxies like our own cannot provide sufficient flux to account for the observed intensity. Most of the distant galaxies are observed out to a red shift of $Z \sim 0.5$. Only a small percentage of the sky is covered by such galaxies up to that distance. If these are to produce the observed diffuse X-ray flux, then the flux from our own Galaxy (if it is typical) would have to be about one hundred times greater than that arriving from outside the Galaxy, as is the case with ordinary starlight. We can thus eliminate the normal galaxies as major contributors to the diffuse X-ray background.

Before we describe the progress made in measurement of the background and the theoretical interpretation of these measurements, we will consider the region of the spectrum below 1 keV. At these low energies there is now clear evidence that a significant part of the diffuse radiation *does* come from our Galaxy. The spectrum at these energies exceeds the intensity expected from a simple extrapolation from higher energies. This can be seen in Fig. 12.1. There is also considerable patchiness at these low energies, a characteristic shared by the distribution of the diffuse clouds of gas and dust which make up the interstellar medium.

The contribution from the Galaxy to the diffuse X-ray background at low energies is thought to originate in clouds of very hot gas having temperatures between 10^5 and 10^6 K. These clouds become heated by collisions between expanding shells of material ejected

DIFFUSE COMPONENT SPECTRUM

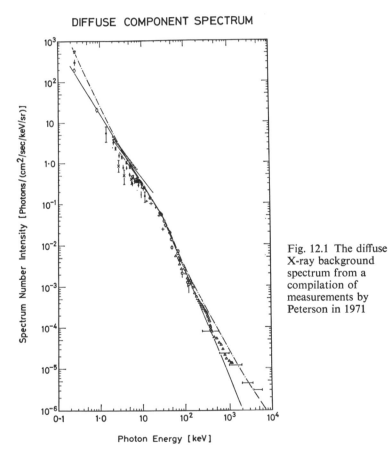

Fig. 12.1 The diffuse X-ray background spectrum from a compilation of measurements by Peterson in 1971

by supernovae explosions. The complex structure thus reflects both the patchiness of the interstellar material itself and the distribution of the supernovae outbursts.

The galactic component at low energies complicates our study of any contribution external to the Galaxy. In particular, it is difficult to detect a very low-density plasma between the galaxies at a similar temperature. Such a plasma could provide enough mass to reduce the expansion of the Universe and eventually cause a collapse to a final singularity. However at higher energies, where the galactic contribution is negligible, measurements and their interpretation can contribute to our understanding of the distant parts of the Universe.

12.2　The theories of extra-galactic X-ray background

The isotropic X-ray background has provided a rich field for theoretical speculation. When the accuracy of the measurements was low the uncertainty in the shape of the spectrum could be accounted for in a remarkable number of theories. Some of these were most ingenious in that they were able to explain details in the spectrum which later proved to be artifacts of measurements. It is sometimes suggested, in jest, that there is a relationship between the accuracy of prediction and of measurement where the product of their errors remains a constant when the measurements become more accurate.

The theories on the origin of the hard X-ray background were reviewed by Joseph Silk in 1970. At one extreme a truly diffuse origin is suggested, where the source permeates the Universe in the form of gas at very high temperatures of 10^9 K. At the other extreme there is the view that the background arises from a large number of discrete sources, which radiate considerably more energy than normal galaxies. The difficulty central to these theories is the form of the spectrum between 1 and 1,000 keV. It is apparent, from the numerous measurements, that the spectrum changes its slope somewhere between 10 and 100 keV. At low energies the intensity changes less rapidly with increasing energy. This change in slope has been difficult to explain in a simple fashion.

One theory proposes the interaction of high-energy electrons with the microwave background. The electrons, according to this theory, provide a source of energy which can raise the energy of the photons in the microwave background to the X-ray region of the spectrum.

The inverse Compton scattering process is an extremely efficient means of conversion into X-radiation. It is assumed that the high-energy electrons leak out from the galaxies. However, the required density of such electrons is in conflict with the observed numbers of high-energy electrons. The quantities of these electrons are, however, measured indirectly through their radio emission in the magnetic fields of radio galaxies, where the synchrotron process operates.

The two main theories which invoke a diffuse origin for the isotropic X-ray background, the inverse Compton process and the theory of hot intergalactic gas, are of interest because it is hard to produce the necessary flux of radiation without placing the source at much earlier stages in the evolution of the Universe. More energy may have been available at times corresponding to red shifts $(Z \sim 5)$. For the higher-energy gamma radiation, even earlier epochs $(Z \sim 100)$ have been invoked in some theories. The background radiation may thus tell us something about the early history of the Universe, between the time when it consisted of radiation and the time of the formation of the galaxies.

A different approach to the problem of the X-ray background has become available with the discoveries of the large number of discrete sources outside our Galaxy. The two types of extra-galactic source, discussed in the previous chapters, can be examined to see whether there are sufficient numbers to provide the measured flux. We can also try to fit the observed spectrum of the diffuse flux by combining the characteristic spectra of the clusters of galaxies and the spectra of the active galaxies. A third technique, also based on measurement, is the investigation of small-scale irregularities in the distribution of the background radiation. These approaches to the study of the background are called superposition theories. The way in which the number of sources in a given intensity-range change with intensity is of fundamental significance in these studies.

12.3 The source–count relationship for extra-galactic sources

In the radio region of the electromagnetic spectrum the study of the extra-galactic sources and of their intensity provided the first evidence for an evolutionary universe. In a steady-state isotropic universe the number of sources $N(S)$ greater than a given intensity (S) would give a relationship

$$N(S) = \text{constant} \times S^{-3}/_2$$

A plot of log N against log S would then give a straight line with a slope of $-3/2$. This relationship is often referred to as the three-halves power law. As radio telescopes became more sensitive, with very large arrays to collect the weak signals, it became possible to test the expected relationship between log N–log S. After some initial confusion and a great deal of controversy, it became clear that the slope was steeper than expected for the brighter sources in the survey. Recent experimental data was obtained by Pooley and Ryle and is shown in Fig. 12.2. The accepted explanation for the steeper slope of the higher luminosity sources is that, in the past, most sources had a higher luminosity or alternatively that there was a greater concentration with increasing distance than exists locally. These conditions are not available in a steady-state theory of the Universe, hence the close scrutiny of the early experimental data. The flattening of the slope of the log N–log S relation as one approaches the fainter sources is interpreted as the result of weakening of the observed radiation through the effects of large red shift of the spectra. It is expected that this effect eventually dominates the evolutionary trends at large red shifts.

With the discovery of extra-galactic X-ray sources, it was natural that astronomers would want to investigate the log N–log S relation

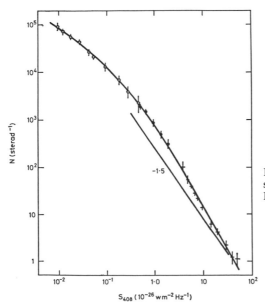

Fig. 12.2 The radio-source counts derived by Pooley and Ryle in 1968

in this part of the electromagnetic spectrum. The estimates of the contribution to the diffuse X-ray background from the clusters of galaxies and the active galaxies fall short of the observed intensity by about a factor of 3. An evolutionary trend similar to the radio sources would help to explain the deficit.

Some progress has been made in the determination of the log N– log S relation for X-ray sources. At the time of writing, the surveys of sources are limited in sensitivity to a strength of one-tenth of the strongest extra-galactic source. This range is barely adequate to discriminate between evolutionary models and a uniform distribution of X-ray sources throughout the Universe. It has, however, been possible to extend the relation to smaller values of the source intensity by investigating the fluctuations in the background, which are observed as an X-ray detector scans the sky. Warwick and his colleagues at the University of Leicester detected such fluctuations in data from the survey instrument on the Ariel V satellite. They showed that the log N–log S relation has a slope consistent with a value $-3/2$ for sources which have a strength down to one-hundredth of the brightest sources. There is thus no clear indication that X-ray sources were more luminous or more numerous at early stages in the evolution of the Universe. It is clear that surveys of much greater sensitivity are needed if we are to make studies comparable with those available for radio sources.

In the United States, Boldt has led a group investigating the X-ray background in the energy range from 1 to 50 keV. Their instrument, flown on the High Energy Astronomical Observatory (HEAO 1), was specially designed to study the background. There is a persistent problem in most investigations of the diffuse X-ray background: how to estimate the part of the signal which is caused by unwanted interference resulting from charged particles. The usual technique involves obscuring the X-ray flux by means of a shutter. The Earth itself has been used as a shutter to obtain the counting-rate in the detector in the absence of X-ray signals. Alternatively, the X-ray signals can be removed by placing heavy metal doors over the detector. Both techniques have inadequacies. It is known that the Earth can itself produce X-ray signals, particularly in the higher energy range greater than about 20 keV. Under these conditions it is difficult to estimate the contribution from the Earth's atmosphere. This can also be highly variable during bombardment by gamma rays and particles from solar flares. The use of obscuring doors, next to the detector, suffers from the disadvantage that the production of background signals from charged particles may also be modified when the doors are closed.

The instrument designed by Boldt and his colleagues determines the flux and spectrum of the X-ray background by using three different beam widths of otherwise identical detectors. The true value for the diffuse background can be obtained by comparing the signals from these detectors. Early results from Boldt's experiment suggest that the spectrum of the X-ray background fits that of a high-temperature plasma (at about 10^9 K) in the energy range up to 40 keV. From then on the spectrum appears to fit smoothly with the established power law. There is thus no evidence for a sharp break in the spectrum between 10 and 100 keV. It is expected that when this data is analysed fully, the spectral fluctuations will also give improved values for the log N–log S relation.

The study of the X-ray background indicates in a striking way how advances in astronomy are made, as instruments of greater sensitivity and better spectral resolution become available. The first true observatory dedicated to X-ray astronomy—the Einstein satellite (see Plate 6) is demonstrating how progress can be made by use of pointed X-ray-reflecting telescopes. The nearest galaxy, M31, which is known to emit X-rays, is now seen to contain more than sixty discrete sources. The observation of this galaxy over lengthy periods will undoubtedly reveal several examples of eclipsing binary sources and the rare occurrence of a black hole in a binary system may also be discovered.

Lengthy observations of regions which had been thought to contain no known sources have shown that many weak sources of X-rays do in fact exist. Some are associated with stars in our Galaxy while others are quasars at immense distances. Eventually we may be able to discover how these powerful sources are distributed throughout the Universe.

Better spectral resolution of the solid-state detector, used for the first time on the Einstein satellite, has begun to reveal the complexities of the spectra of the supernova remnants and the binary X-ray sources. Lines that are characteristic of sulphur, silicon and several other elements have been clearly resolved.

All of the clusters of galaxies observed by the Einstein telescope are copious emitters of diffuse X-rays. The X-ray-wavelength range is thus becoming as central to astronomy as the restricted visible region of the spectrum.

12.4   The future

X-ray astronomy has now reached a degree of maturity where the immediate needs for further development can be outlined with some confidence. We can expect that very sensitive surveys of the extra-galactic sources will give us an insight into the characteristics of sources at the early stage in the evolving Universe. The instruments for the detailed study of the nearby sources can also be fairly readily defined, and we can expect large arrays of detectors to measure the shortest time scales of the fluctuations in the accretion discs around compact objects. There will be high-resolution spectrometers to measure the characteristics of the plasmas, which exist in clusters of galaxies, in the remnants of supernovae and in the regions around compact objects. There is a need to monitor the whole of the celestial sphere to detect the transient sources and the rapid bursts of X-rays whose origin is still unknown. The signals from some X-ray sources may also be strongly polarized. Measurements of the polarization may give insight into the mechanisms of the accretion discs surrounding black holes. In fact, there is no lack of ideas on what needs to be studied or the means by which the measurements can be made. What is lacking is the allocation of the resources to continue the subject at even a modest level.

There is at present a crisis of resources. This can be seen immediately if one looks at the diminishing budget of the United States' space activities carried out by NASA. In Fig. 12.3 the percentage of gross domestic product devoted to space activities is shown for the years 1968 to 1975. It is clear that over this period the USA, which has provided opportunities for space investigations in a most generous manner to its own scientists and those of many other countries, has suffered a halving of the budget.

During the 1960s, and the first part of the 1970s, space astronomy was a very small part of the total space activity. The Apollo programme of lunar landings and the follow-on Skylab programme were immense undertakings compared with the instrumented unmanned satellites, which opened the ultraviolet and X-ray windows to the stars.

The problem for space astronomers can be simply stated. Unlike many other scientific pursuits, the information from the heavens comes by a one-way channel. We can never hope to modify the distant parts of the Universe as a result of the signals we receive. The information we obtain enhances our culture through its aesthetic

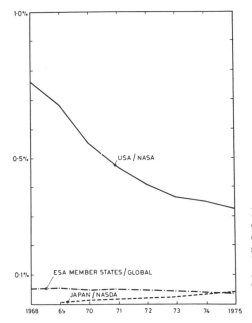

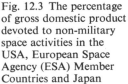

Fig. 12.3 The percentage of gross domestic product devoted to non-military space activities in the USA, European Space Agency (ESA) Member Countries and Japan

value. In a world which always looks for a more tangible return on its investment, this presents the astronomer with a serious problem. Galileo faced a similar difficulty when pioneering the telescope in the early seventeenth century, as can be seen in the following passage:*

> On 24 August 1609 Galileo presented the Venetian government with an improved version of his instrument [telescope] . . . explaining how valuable it would be: 'We could discover the enemy ships more than two hours sooner than he could discover us'. Next day the Doge, ruler of Venice, and his council expressed their appreciation by confirming Galileo in his post for life at a salary of a thousand florins.

The modern astronomer can also demonstrate the advantages of using his techniques in other pursuits but he needs a great deal more support than a major city can provide. The largest telescopes on the ground are now usually financed by individual governments and have become legendary. The space telescope to be launched in the mid-1980s is supported by both Europe and the United States. It will appeal as a space spectacular and may well capture

* Vincent Cronin, *The Flowering of the Renaissance* (Collins, 1969).

imagination and support in much the same way. Plans for a large X-ray observatory are also well advanced and an early concept can be seen in Fig. 12.4. The mirror for this telescope will provide resolving power which may exceed that of the space telescope itself.

The pessimists see no possibility of maintaining support for space astronomy without appealing to the spectacular, both in performance and cost. The disadvantage of placing limited resources in a single space telescope for the entire community is clear: the observing time will be greatly over-subscribed.

There is, however, some room for optimism. The high cost of space astronomy (a modest optical telescope in orbit costs about double that of the largest telescope on the ground), may be reduced in the next decade. The space transportation system under development by NASA aims to provide lower costs for launching satellites into orbit. Its ability to rendezvous with satellites or space platforms

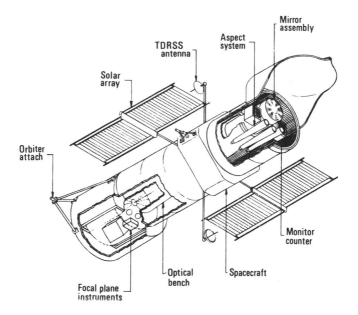

Fig. 12.4 The advanced X-ray astrophysics facility, a free-flying observatory that is Shuttle-launched, maintainable on-orbit, and retrievable. AXAF is conceived as a 1.2 m diameter, 10 m focal length X-ray telescope with interchangeable focal plane instrumentation. It should provide astronomers with the means of observing all known visible objects in the X-ray part of the spectrum

may mean that space astronomy can be carried out as cheaply as the provision of major ground-based observatories. This may transform our attitudes to the use of space platforms. Studies of a refurbishable satellite system have indicated that considerable savings can be made if instruments are exchanged and carried in orbit by means of NASA's space shuttle and its remote manipulating systems. In Fig. 12.5 a satellite-platform concept is shown whilst being retrieved by the space shuttle prior to the exchange of its instruments. A long-term programme of space astronomy may eventually become as efficient as the provision of observatories on high mountains.

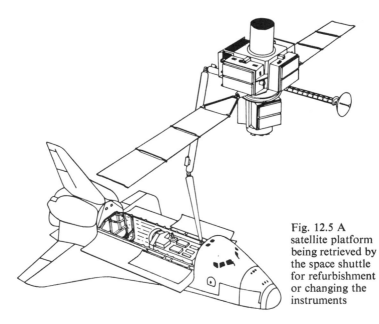

Fig. 12.5 A satellite platform being retrieved by the space shuttle for refurbishment or changing the instruments

If we wish to pursue astronomy without having to provide space spectacles or appealing to the spin-off available to industrial or military activities, we must ensure that the aesthetic rewards of astronomy are available to all. The data must be carefully processed and the concepts of the invisible astronomy clearly presented. Our aim in this book has been to help improve the perception of astronomy, and we may well have encountered more problems than solutions. However, questions of origin and dissolution are woven into most cultures and we are now beginning to be provided with better means of looking to the heavens for our answers.

Bibliography

The following books are useful references for readers who may wish to study subjects in more depth than has been possible in this book:

BOYD, R. L. F., SEATON, M. J., and MASSEY, H. S. W., *Rocket Exploration of the Upper Atmosphere*, Pergamon Press, Oxford, 1954.

COMPTON, A., and ALLISON, S., *X-rays in Theory and Experiment*, Macmillan, Basingstoke, 1935.

GIACCONI, R., and GURSKY, H., *X-ray Astronomy*, D. Reidel, Dordrecht, 1974.

GURSKY, H. and RUFFINI, R., *Neutron Stars, Black Holes and Binary X-ray Sources*, D. Reidel, Dordrecht, 1975.

MITTON, S., *Exploring the Galaxies*, Faber and Faber, London, 1976.

MITTON, S., *The Crab Nebula*, Faber and Faber, London, 1979.

Readings from Scientific American: Frontiers in Astronomy, W. H. Freeman, New York, 1970.

Readings from Scientific American: New Frontiers in Astronomy, W. H. Freeman, New York, 1970.

SANFORD, P. W., and LASKARIDES, P. (eds.), *Galactic X-ray Sources*, J. Wiley, Chichester, 1981.

SCIAMA, D., *Modern Cosmology*, Cambridge University Press, Cambridge, 1971.

UNSOLD, A., *The New Cosmos*, Longman/Springer Verlag, Harlow, 1969.

Index